T0329422

Advances in Analog and RF IC Design for Wireless Communication Systems

Advances in Analog and RF IC Design for Wireless Communication Systems

Gabriele Manganaro

Domine Leenaerts

AMSTERDAM • BOSTON • HEIDELBERG • LONDON
NEW YORK • OXFORD • PARIS • SAN DIEGO
SAN FRANCISCO • SINGAPORE • SYDNEY • TOKYO

Academic Press is an Imprint of Elsevier

Academic Press is an imprint of Elsevier
The Boulevard, Langford Lane, Kidlington, Oxford OX5 1GB, UK
225 Wyman Street, Waltham, MA 02451, USA

First edition 2013

Notice
No responsibility is assumed by the publisher for any injury and/or damage to persons or
property as a matter of products' liability, negligence or otherwise, or from any use or operation
of any methods, products, instructions or ideas contained in the material herein. Because of
rapid advances in the medical sciences, in particular, independent verification of diagnoses and
drug dosages should be made.

Library of Congress Cataloging-in-Publication Data
A catalog record for this book is availabe from the Library of Congress

British Library Cataloguing in Publication Data
A catalogue record for this book is available from the British Library

ISBN: 978-0-12-398326-8

For information on all Academic Press publications
visit our web site at books.elsevier.com

Printed and bound in the US
13 14 15 16 17   10 9 8 7 6 5 4 3 2 1

Working together
to grow libraries in
developing countries

www.elsevier.com • www.bookaid.org

Contents

Wireless Infrastructure

1

Gabriele Manganaro[a] and Domine Leenaerts[b]

[a]*Analog Devices, Boston, USA*
[b]*NXP Semiconductors, Eindhoven, The Netherlands*

INTRODUCTION

Wireless communication is today one of the most important ways to transport voice, video, and data using radio-frequency (RF) or microwaves. In fact, since 2002, more phone calls are made via a wireless link rather than a wired link. Even more striking, according to recent statistics on a global scale, today there are already more mobile phone subscriptions than people with access to electricity or access to safe drinking water. The modern mobile handheld terminal, the smartphone, is a clear example of it; it transports voice via a GSM, or W-CDMA cellular communication pipe and data/video can be transported via the WLAN connectivity pipe. Microwaves are typically used for satellite links to broadcast television, to enable internet for remote areas, and for private networks.

In a wireless infrastructure, wireless devices can communicate with each other or communicate with a wired network. In a cellular wireless infrastructure the communication goes first through an access point, called the base station. The mobile handheld device communicates via the base station (BTS) with another mobile device or can access the wired network. Communication takes place in the RF cellular frequency bands from 700 MHz up to 3 GHz with the upcoming LTE-A operating in even higher frequency bands between 3 and 4 GHz. Base stations communicate between themselves via a wired (optical) or again wireless link. The wireless link is very important and is referred to as the microwave backhaul of the base station. The communication happens at various microwave frequencies in the 13–40 GHz range. As base stations do have a fixed geographical position, the communication between base stations is a so-called point-to-point communication, the antennas needed for the microwave backhaul are pointed to each other (i.e. line-of-sight setting).

A satellite (wireless) infrastructure operates in a similar way, where the satellite can be considered as the base station or the access point. At the other end one finds the indoor-unit or set-top-box which is connected to the outdoor unit, the transceiver, and the dish. Most often the wireless downlink communication happens in the 10.7–12.75 GHz Ku-band, but the 18.2–22.0 GHz Ka-band is becoming popular too.

Manganaro: Advances in Analog and RF IC Design. http://dx.doi.org/10.1016/B978-0-12-398326-8.00001-7

Uplink communication, that is the link from earth to satellite, takes place in the 13 GHz or 30 GHz frequency bands.

1.1 The cellular infrastructure

Mobile data traffic is expected to increase 18-fold over the next five years to 11 Exabytes per month by 2016 as a result of an ever-increasing demand for data throughput driven by widening spread of smartphones and, even faster, of tablet devices and other data-hungry mobile devices. In the meantime, wireless technologies like LTE Advanced are approaching the physical limits for achievable channel capacity because the spectrum is a limited resource. The consequence of all this is that further growth of the network capacity must come from new networks where for instance macro cells are overlaid by small cells and offload of mobile networks might take place with seamlessly integrated WiFi access points.

A macro cell base station delivers the best performance and coverage, but is very expensive to roll out. The cell radius of a macro base station is around 1–25 km and can handle more than 256 users. The average transmitted power is more than 10 W; peak power is more than 100 W. A macro base station consists of one or more reasonable-sized cabinets plus a big tower, which means that in very populated areas acquiring a site to install the macro base station might be difficult and very expensive.

The small cell can densify the macro cell network in urban areas with a lower total cost of ownership. Small cells, a collection of pico and micro cells, typically have a cell size between 200 m and 1 km and can handle up to 256 users. The average transmit power is around 5 W. However, a high-performance backhaul is mandatory for optimal performance. This backhaul can be wired or wireless. The macro cells and small cells together should deliver a high-quality (e.g. best coverage and capacity) user experience for voice and data.

The addition of WiFi access points should improve the capacity in high-traffic areas and "hot spots" and is mainly data centric.

The backhaul is a key element for high-performance cellular networks. The majority of the microwave, wireless, links use the spectrum between 6 and 38 GHz and offer capacities up to 400 Mbps with 56 MHz channel bandwidth and 256 QAM constellation. The main challenge in the microwave backhaul is the required line-of-sight (LOS) and strict alignment between the two end points. The requirement of LOS can be relaxed by the introduction of beam forming techniques and/or advanced MIMO techniques.

1.2 The satellite infrastructure

The microwave satellite market is a mass market with over 80 million outdoor units in 2012. The majority of the outdoor units are for satellite TV reception in the Ku-band. This is a mainly one-way communication pipe with the satellite. A smaller market is the two-way communication pipe for applications like credit card data, internet in remote areas, maritime communication, and private networks (e.g. CNN).

This very-small-aperture-terminal (VSAT) communication uses data rates up to 4 Mbps. The name VSAT depicts the size of the antenna, a dish antenna that is smaller than 3 m.

The low-noise block (LNB) outdoor unit is the receiver converting the received satellite signals in the Ku-band down to the L-band (950–2100 MHz). The LNB is connected via a coax cable to the set-top-box in the house, where the costumer can select the TV channel. The set-top-box will power the LNB via the coax cable. Consequently, as the set-top-box can only deliver a certain amount of power, the power consumption of the LNB is a critical design parameter.

1.3 Challenges

Wireless infrastructure, be it cellular or satellite, is a more professional market segment rather than consumer market segment. Consequently, the market is traditionally driven by performance demand rather than by cost. A typical RF card in a base station or a low-noise block down converter unit in a satellite receiver is mainly based on discrete III–V compound technology components. They deliver the required performance, while the associated cost and power consumption is a lesser issue. But this landscape is changing. Mobile phone operators see their energy bills exploding while they are expanding the capacity of their cellular infrastructure. And as the expansion is taking place in dense areas like shopping malls, the installed access point should be small in size too. If satellite communication should become available in developing areas such as India and South America, a price reduction would be needed.

Consequently, there is a need to move away from III–V compound technologies toward silicon-based solutions, which inherently can provide cost reduction and power dissipation reduction, similar to what has been observed in mobile handheld devices in the past. However, the required performance, more easily delivered by GaAs-like technologies, should not be compromised. This is a challenge for silicon-based technologies where the active devices inherently have poorer performance. The RF radio especially will face challenges from this technology change.

This book will address the various issues related to the transition from III–V compound technology toward silicon-based solutions for RF radios in wireless infrastructure.

On the other hand, the data converters (ADCs and DACs) in the same signal chain have already seen a substantial shift from a mix of BiCMOS and CMOS designs to a near monopoly of CMOS data converters in the most recent designs. Some of the driving forces behind this technology shift follow.

First of all, starting from 0.18 µm and even more with finer-lithography CMOS processes, the transition frequency, f_T, of MOS devices has become very competitive with the one of many of the bipolar devices available with some of the mainstream BiCMOS processes and for a comparable price. Moreover, both the active and passive device matching have been steadily improving with CMOS scaling. Combining that with the far greater level of integration achievable in CMOS makes nanometer-scale CMOS processes far more attractive than BiCMOS alternatives in designs with moderate resolution (10–16 bits) and a high sample rate (100 MSPS–1 GSPS and beyond) that are needed for wireless infrastructure.

One of the many design challenges associated with this process technology shift is represented by the lower and lower power supply usable with scaled CMOS processes. Specifically, 3.3 V and 5 V analog supplies, which were common for older BiCMOS data converters, have been replaced with 3.3 V, 2.4 V, 1.8 V, and, more recently, 1 V and even 0.9 V analog supplies in deep nanometer CMOS data converters as 40 nm or 28 nm. This has inevitably led to formidable design challenges due to lowering voltage headroom and the ability to design analog circuits with a sufficiently low noise power spectral density with rapidly shrinking usable signal power.

The greater adoption of CMOS processes for high-performance data converters for wireless infrastructure has also fueled another trend. Scaled CMOS processes have enabled dramatic improvements in terms of both cost- and power-effective digital signal processing functionality. Therefore it is very common in present-day wireless infrastructure data converters that a good deal of the digital post-processing in the receive path is integrated on the same die with the ADC and, conversely, that the digital pre-processing in the transmit path is integrated on the same die with the DAC. Such digital functions include, for example, digital down conversion, filtering, channel separation in the case of the receive path; and consist, for example, of digital up conversion, interpolation, filtering, etc., in the case of the transmit path.

Furthermore, this on-chip availability of efficient digital functionality has also enabled the practical implementation of on-chip calibration and nonlinear correction schemes for some of the cited analog-domain shortcomings. So, in a way, the source of some of the analog design grief introduced by nanometer processes can also be tackled by means of the rich digital processing power introduced by Moore's law by a broad new slew of techniques loosely categorized as "digitally assisted analog (and RF) design."

Furthermore, the increasing demand for larger data throughput, combined with the increasing availability of computational DSP power fueled by Moore's law, has driven the demand from BTS manufacturers for wider and wider bandwidth signal chains able to process co-existing and different communication standards (not only GSM channels, but also LTE, WCDMA, etc.). Without doubt, that has made the RF front-end and the data converters the "performance bottleneck" of this class of radio systems. Along with that, the power, functionality, and cost advantages of DSPs have fueled the quest for the "holy grail" of radios, known as "software radio" and consisting of a signal chain where the RF and analog front-end are increasingly smaller and where the analog-to-digital conversion and the digital-to-analog conversion boundaries move closer and closer to the antenna, allowing more of the (de-)modulation/filtering/processing of the communication channels to be efficiently and reliably performed in the digital domain. Indeed one of the chief challenges of this quest lies in the fact that moving the boundary between the physical analog medium and the digital one toward higher frequency or wider bandwidth is paid for by dramatic increases in power consumption to perform the conversion.

It is mainly because of this reason that for a given set of communication specifications and development and implementation costs a balanced choice for the borderline between analog and digital leads to sensitive signal-chain trade-offs.

In many commercial cell phone BTSs this boundary lies presently between the tens and hundreds of MHz and where the ADC samples the lowest intermediate frequency (IF) stage of a heterodyne scheme or where the DAC directly synthesizes the signals at one of the IF frequencies, or possibly at RF.

Another critical design challenge originates from the need, mentioned above, to develop a BTS that is physically small in size: going from a unit similar in size to a small refrigerator to one similar to a desktop PC or possibly slightly bigger than a laptop. To reduce physical size, higher integration is certainly one lever, but most importantly heat removal is a very important challenge. Heating sinks, fans, and other sizeable devices for heat removal and management need to be eliminated. That means the electronics need to dissipate less heat and need to be able to operate in a much hotter ambient temperature. Clearly the previously cited power consumption challenge originating from increasing communication performance demands is further aggravated by these much more challenging operating environment conditions. Not only circuit design technology and silicon reliability are challenged. Electrical, thermal, and mechanical aspects of package technology, assembly and manufacturing, printed circuit boards, to cite a few, come into play and trade-offs between all these aspects take center stage in the design development process.

1.4 **This book**

This book aims to provide an overview of the main and most current technical topics in radio-frequency and analog/mixed-signal IC design for wireless infrastructure. The chapters are contributed by some of the most well-respected professionals in this field both from industry as well as academia.

The chapter authored by H. Darabi provides a broad overview of the architectures and system trade-offs involved in the overall design of CMOS transceivers. Following that, the individual chapters dive into covering the challenges and the design techniques associated with all the main functional blocks in modern commercial cellular base station systems. Following an order that somewhat mirrors the cascade of stages between the antenna and the DSP, we begin with a chapter on the design of low-noise, high-linearity amplifiers (LNAs) authored by D. Leenaerts as well as another chapter on RF power amplifiers written by M. Acar et al.

That is followed by a chapter on frequency synthesizers/PLLs contributed by S. Levantino and C. Samori describing low phase noise oscillators used in down and up conversion. And, of course, coupled with that is a chapter on mixers and modulators by W. Redman-White. The special case of low noise down-converters for satellite communication systems is covered in a chapter by P. Philippe et al. The latter concludes the part of this book on what is commonly considered as the true radio-frequency electronics of this class of infrastructure radios.

The many classes of analog-to-digital and digital-to-analog converters found in base stations are discussed in the chapters that follow. R. Schreier and H. Shibata authored a chapter on emerging continuous time band-pass $\Delta\Sigma$ ADCs for communication applications. A more established ADC architecture in this arena is the pipelined ADC. That is the topic of the following chapter by M. Elliott and B. Murmann. Another ADC architecture that is gradually emerging as a viable option is the SAR ADC. Interleaving multiple lower-sampling-rate ADCs allows the building of overall higher-sampling frequency converters as described in the chapter by K. Doris et al. that follows.

Finally a chapter by G. Engel and G. Manganaro describes the other side of converters, namely modern high-performance digital-to-analog converters for the transmit path.

Last but not least, the chapter on time-to-digital conversion for digital frequency synthesizers describes an emerging breakthrough alternative to some of the other approaches in frequency synthesis as well as domain conversion. This chapter is authored by M. Perrott and concludes the volume.

CONCLUSIVE REMARKS

Concluding this introduction, the two editors of this volume would like to acknowledge and thank the many authors of this book, who, on rather short notice and on a fast-paced schedule, have graciously managed to contribute a comprehensive and cohesive set of outstanding chapters on very critical topics for a very rapidly evolving and highly competitive space. Given the fast pace of innovation and the breadth of this field there is no doubt that reading this material will both answer questions and trigger new ones. In fact, the contents are very much on the tipping point of today's state-of-the-art and knowing what the right questions and the challenges are is as important as being able to tackle them.

The editors would also like to thank the anonymous reviewers. Providing us with their constructive criticism and suggestions has allowed us to improve the organization and blend of the topics, hence helping us to structure the many moving parts of this volume into a more organic and thorough treatise of the topic.

Our thanks also go to the staff at Elsevier and, in particular, to the senior commissioning editor Tim Pitts. Tim, with his British gentleman style, his instinctive flair, and his punctual prodding, was able to first convince us to embark on this adventure and then to actually deliver it on a schedule that is sensitive to the timeliness of its contents.

Last but not least, we thank our families for their unconditional support and encouragements, despite the precious time taken away from them after coming back from already long and tiring days at our regular day jobs.

CMOS Transceivers for Modern Cellular Terminals

2

Hooman Darabi

Broadcom Corporation, Irvine, CA, USA

INTRODUCTION

The cellular handset market is expected to ship a total worldwide volume in excess of 1.2 billion mobile phones with semiconductor content valued at close to $20 billion dollars per year in 2013 [1]. The handset market segment that is leading the growth for the entire handset industry is the ultra-low-cost segment. The CAGR (cumulative annualized growth rate) for this segment is 58% (2006–2014) which is significantly larger than the smartphone segment of 18% CAGR (2006–2014), the next largest segment, with all other segments facing negative growth rates [2]. Similar to most other standards, the main drive behind the cellular growth arises from the demand for higher data rates while maintaining the mobility as shown in Figure 2.1. This has introduced several new additions to the original 3G standard introduced in 2001, such as HSPA, HSPA+, or diversity to support higher throughput, and ultimately the LTE, which can achieve up to 300 Mbps data rate for the downlink. Unlike local area network applications (LAN) where mobility is not a concern, the cellular handset can support mobility of up to 100 km/h.

This book chapter focuses on modern cellular handsets and provides an overview of the standard with detailed discussion on the requirements of the transceiver. The key radio requirements are derived, and translated to circuit specs, giving an overview of a practical top-down radio design. Our goal is to provide a concise yet complete analysis of RF system requirements of 2/3/4G handsets and offer guidelines as to how derive the circuit level specifications for almost all the key blocks of modern cellular transceivers.

This chapter is organized as follows: We will discuss GSM/EDGE requirements in Section 2.1, then focus on 3G and LTE transceivers in Section 2.2. In Section 2.3 we discuss handset calibration techniques, followed by conclusions.

2.1 2G transceiver architecture design

In this section we will consider the system requirements of 2G (GSM/EDGE) transceivers. GSM/EDGE is a TDMA (time division multiple access) system with four bands of operation, listed in Table 2.1. Each channel is 200 kHz wide, with RX and

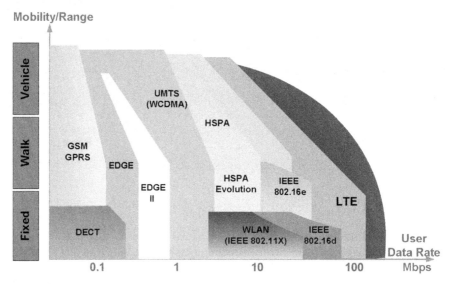

FIGURE 2.1

Evolution of wireless devices.

Table 2.1 GSM/EDGE Bands of Operation

Band	GSM 850	E-GSM	DCS 1800	PCS 1900
RX Frequency, MHz	869–864	925–960	1805–1880	1930–1990
TX Frequency, MHz	824–849	880–915	1710–1785	1850–1910
Bandwidth, MHz	25	35	75	60

TX at two separate bands. GSM supports up to 270 kbps data rate using GMSK modulation, while EDGE is three times faster using 8PSK modulation scheme.

2.1.1 GSM/EDGE receiver requirement

Most modern 2G receivers use either zero or very low-IF architecture. A generic block diagram of such a receiver is shown in Figure 2.2, consisting of the front-end (LNA and mixer), channel-select filter, and the ADC. The LNA is typically preceded by a SAW filter (to eliminate out-of-band blockers) and a switch (to support TDMA).

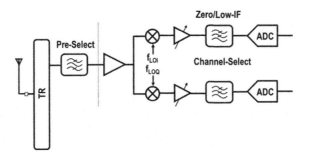

FIGURE 2.2

A generic zero- or low-IF receiver.

To illustrate the GSM blocker challenges, Figure 2.3 shows an example of the out-of-band blocker profile for the PCS band. According to the 3GPP standard [3–5], a blocker as large as 0dB and as close as 80MHz to the edge of the band could accompany a weak desirable signal only 3dB above the reference sensitivity. The other three bands are similar, although for the two lowbands, the minimum spacing is 20MHz.

Given that the desirable signal is weak, the LNA gain is still high, and the 0dBm blocker can heavily desensitize the RX front-end. For these reasons it is common to place an external SAW filter at the LNA input to reduce the blocker level to the point that the receiver can handle (at least to −23dBm, which is the highest in-band blocker described below), although there are a few recent attempts that have solved this issue by employing integrated high-Q filtering [14–19,33] and linearizing the RX [34].

In addition to blocker tolerance, the sensitivity is a basic requirement of any application, and particularly the GSM. According to the 3GPP standard, a voice sensitivity of −102dBm is needed, although most advanced phones target a much better requirement of −110dBm. Assuming 3dB front-end loss (the SAW filter and switch), a modem signal-to-noise ratio (SNR) of 5.5dB, and a data bandwidth of 200kHz, the required radio noise figure is:

$$\text{NF} = 174 + \text{Sensitivity} - \text{SNR} - 10 \times \log(\text{BW}) - \text{Loss} = 2.5\,\text{dB}, \quad (2.1)$$

which is the target of most modern designs.

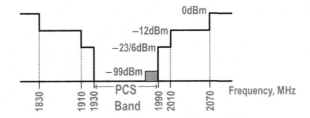

FIGURE 2.3

GSM out-of-band blockers profile for PCS band.

Another challenge in GSM receivers arises from the in-band blocking requirements (Figure 2.4), and particularly the 3 MHz blocker which is −23/−26 dBm for LB/HB.

This is described as follows:

1. Since the desired signal is only 3 dB above the reference sensitivity of −102 dBm, the front-end gain is at or very close to the maximum. Even though the relaxation in the sensitivity potentially allows for some RF gain reduction, in practice this is very limited, as −99 dBm level coincides with some other key data sensitivity tests such as MCS8 or MCS9. This leaves little room for SNR degradation caused by the potential RF gain reduction, and thus the −23 dBm blocker sets the receiver compression point at the maximum front-end gain. Since the in-band blockers are specified to be static sine waves, some compression gain may be acceptable as long as the receiver noise figure is not compromised much.
2. The reciprocal mixing imposed by a 3 MHz blocker sets the LO chain phase noise. It can be easily shown that assuming a phase noise value of PN at the blocker offset, and a blocker level of P_B, the receiver noise figure with everything else ideal is equal to:

$$BNF = 174 + P_B + PN \qquad (2.2)$$

For instance with a HB phase noise of −140 dBc/Hz at 3 MHz, and a blocker level of −26 dBm, the receiver noise figure due to the reciprocal mixing alone is $174 − 26 − 140 = 8$ dB.

Assuming a thermal noise figure of 5.5 dB at maximum gain (Eq. (2.2), with 2.5 dB radio NF and 3 dB loss at front-end), the overall receiver noise figure in the presence of the 3 MHz blocker is 9.9 dB. With the desired signal at −99 dBm, according to Eq. (2.2) the total blocker noise figure (BNF) allowed can be as high as 13.5 dB, leaving about 3.5 dB of margin for PVT variations. This assumes that the receiver suffers little compression.

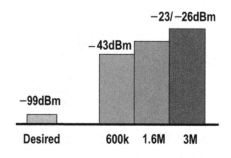

FIGURE 2.4

GSM in-band blockers profile.

3. The receiver selectivity is another concern. For the case of the 3 MHz blocker, the IF chain including the analog filter usually does not suffer from any compression as the blocker is far away, but the RF chain output and particularly the mixer buffers need to enjoy enough filtering to minimize the compression.

In addition to in-band blockers, GSM defines a very stringent adjacent blocker profile as shown in Figure 2.5. The most challenging requirement is usually set by the 400 kHz adjacent blocker, which is 41 dB higher than the desirable signal at −82 dBm. In the case of a low-IF receiver, assuming low-side injection for instance, the blocker at −400 kHz (low-side) is the most problematic one, appearing at a distance of 400 kHz − 2 × IF after the down-conversion, which could be very close to, or overlapping the desirable signal. The channel-select filter will help very little, which imposes a very tight requirement on the ADC as well as the receiver image rejection.

The higher the IF, the closer the blocker, and thus more stringent image rejection is needed, as illustrated in Figure 2.5. In the extreme case of 200 kHz IF, the image, that is the −400 kHz blocker, downconverts entirely on the desired signal and image rejection (IMRR) is thus −41 dB − SNR = −50 dBc, assuming an SNR of 9 dB for the modem (5.5 dB plus some margin for the other contributors). For these reasons most current receivers use a very low IF of around 100–150 kHz, requiring an IMRR of about 40 dB, which is done almost entirely in digital domain post-ADC. This, however, sets a very large dynamic range requirement on the ADC itself, so as not to be saturated by the −400 kHz blocker (which is subject to almost no IF filtering). For instance if the ADC quantization noise is set to be 30 dB below the signal (so as not to degrade the receiver noise figure noticeably, assuming 20 dB room for down-fading), an up-fading target of, and assuming the blocker stays at least 3 dB

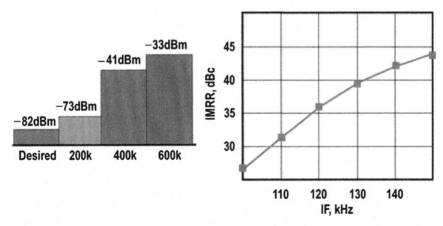

FIGURE 2.5

GSM adjacent blocker profile and required image rejection.

below the ADC full scale (to leave room for AGC error), the ADC dynamic range is: $30 + 41 + 3 + 15 = 89\,\text{dB}$. This is graphically shown in Figure 2.6.

Note that the signal and the blocker are subject to the same level of amplifications as they progress through the receiver, given that the blocker is subject to very little IF filtering. Moreover, the up-down-fading is caused by the fast fluctuations in the signal and adjacent blocker levels that cannot be tracked by the AGC. The phase noise requirements due to reciprocal mixing of the adjacent blockers, which is typically relaxed and usually meets the 600 kHz and 3 MHz in-band blockers, will automatically satisfy the requirement, assuming a normal 20–30 dB/Dec phase noise profile.

Finally, the IIP3 and IIP2 requirements are calculated as follows: In the case of IIP3, two tones are specified at 800 kHz and 1600 kHz from the desirable signal at $-99\,\text{dBm}$ and at $-49\,\text{dBm}$. Given that: $2 \times \text{IIP3} = 3 \times P_B - \text{Sensitivity} + \text{SNR}$, where P_B is the blocker power, in this case the two tones at $-49\,\text{dBm}$, sensitivity is $-99\,\text{dBm}$, and a SNR of 9 dB is assumed for the modem, the IIP3 is found to be $-19.5\,\text{dBm}$. This is relatively an easy target to meet; however, the 3 MHz blocker will set a more stringent requirement on the receiver linearity and thus IIP3. For example assuming a 1 dB compression of $-23\,\text{dBm}$ to tolerate the 3 MHz blocker, the IIP3 is calculated to $-13\,\text{dBm}$, assuming $\text{IIP3} = P_{1\text{dB}} + 10\,\text{dB}$, and that the IIP3 is mainly limited by the front-end and not the IF filters (and thus the same IIP3 for 3 MHz and 0.8/1.6 MHz offsets). Similarly, as $\text{IIP2} = 2 \times P_B - \text{Sensitivity} + \text{SNR}$, for a blocker at 6 MHz away at $-31\,\text{dBm}$ level, the IIP2 is $+46\,\text{dBm}$, which is a challenging number to meet. A higher IF relaxes the IIP2 as the blocker falls at DC and is subject to more baseband filtering, but the image rejection needed is worse.

Apart from the sensitivity and blocking requirements, the receiver must be able to achieve a high SNR in the absence of the blockers and with a large desirable signal present at the antenna. This is particularly very important in the EDGE mode of operation, and especially the higher data rates such as MCS9 mode of operation. An example of a typical receiver SNR measured at the modem vs. the LNA input signal level is shown in Figure 2.7 [23]. At low inputs, the SNR is merely limited by the receiver thermal noise

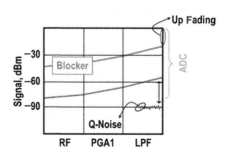

FIGURE 2.6

GSM ADC dynamic range calculation.

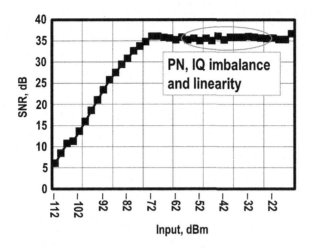

FIGURE 2.7

Typical GSM/EDGE receiver SNR.

figure as stated in Eq. (2.1). In this example, 5.5 dB SNR requires an input level of −112 dBm, translating a sensitivity of −109 dBm at the antenna, assuming 3 dB front-end loss. Assuming the NF degrades at a slower rate as the signal increases (implying that most of the gain reduction is at the IF and not the front-end, thus maintaining a low NF), the SNR increases, until it peaks at a level mostly determined by the receiver in-band phase noise and the IQ imbalance, and to a lesser extent its in-band linearity. For example, an integrated in-band phase noise of −40 dBc (= $0.01 \times 180/\pi = 0.57°$RMS, where $0.01 = -40$dBc is the RMS noise in Radian) limits the SNR to about 40 dB. For 200 kHz GSM/EDGE bandwidth, this approximately corresponds to a stringent in-band phase noise of $-40 - 10 \times \log(200\,\text{kHz}) = -93$ dBc/Hz. Therefore, the 2G receiver in-band phase noise is as important as the far-out noise which as discussed previously resulted from the blocker requirements. For the case of MCS9, the required modem SNR is about 19 dB, translating to a sensitivity of −89 dBm at the antenna for this particular receiver [23].

Note that the receiver has no prior knowledge of whether it is operating in the GSM or EDGE mode, and thus typically one gain table is used between GSM and EDGE. Thus, as pointed out earlier, relaxing the gain for some of the stringent GSM blocker scenarios may potentially hurt the EDGE mode SNR and is usually undesirable.

2.1.2 GSM/EDGE transmitter requirement

Similar to the receiver, the transmitter should meet several requirements to ensure a reliable communication link for the desirable signal (output power vs. sensitivity, phase error/EVM vs. RX SNR) and low interference for the other handsets (mask and receive-band noise vs. in- and out-of-band blockers).

2.1.2.1 *Power and mask requirements*

GSM lowbands require a maximum output power of +33 dBm (Power Control Level 5, or PCL45) for most applications, although up to 39 dBm it is supported by the standard. The power accuracy requirement is ±3 dB, and a total of 15 power levels are supported in steps of 2 dB with ±1.5 dB of accuracy. Thus the minimum power at the antenna is 5 dBm. Given that GSM has a constant envelope modulation, the power amplifier used is typically saturated, and the power control is usually accomplished within the PA through an on-chip power control level. The high bands are the same, except for the maximum required power being 30 dBm (PCL0), which goes down to 0 dBm (PCL15).

The GSM/EDGE mask requirement is shown in Figure 2.8. For GMSK, the most stringent point is 400 kHz, which is required to be −60 dBc, whereas for 8PSK (or 16QAM for EDGE evolution), the mask is relaxed to −54 dBc. The far-out mask (6 MHz and above) is −77/−79 dBc for LB/HB, which is usually met if the 400 kHz is satisfied as far as the phase noise is concerned. For die-band noise level, an absolute requirement of −36 dBm for offsets <600 kHz, and −46/−51 dBm for >1.8 MHz for LB/HB is required as an exception.

Since GSM signal is only phase modulated, the mask is typically set by the phase noise performance of the TX PLL. The critical 400 kHz mask requirement usually dictates key PLL design choices. According to 3GPP 400 kHz mask is measured with 30 kHz RBW (resolution bandwidth), which leads to the following phase noise requirement:

$$PN = -60 - 9 - 10\log_{10}(30 \text{ kHz}) = -113.8 \text{ dBc/Hz}, \qquad (2.3)$$

where the 9 dB ($\approx 10 \times \log(200 \text{kHz}/30 \text{kHz})$) is an approximate adjustment of the reference power when measured with 30 kHz bandwidth compared to the actual signal power, which is roughly 200 kHz wide. To have sufficient production margin with good yield, a target of better than −65 dBc typical spectrum, and −63 dBc worst case is chosen for most designs, leading to a typical phase noise requirement of better than −119 dBc at 400 kHz for the entire PLL. Note that for HB with an output power of 30 dBm at the antenna, the −36 dBm absolute requirement (measured at 30 kHz RBW) results in a mask of roughly 30 dBm − (−36 dBm) − 9 dB ≈ −57 dBc, although

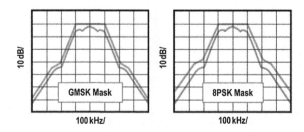

FIGURE 2.8

GSM/EDGE TX mask.

3GPP requires the transmitter to meet the more stringent -60 dBc requirement while -36 dBm is considered an exception.

2G transmitters must be multi-mode devices capable of handling both the constant envelope signals of GMSK modulation in GSM mode of operation as well as time-varying envelope signals of 8PSK modulation in EDGE mode of operation. Despite the fact that the direct-conversion architecture [6] is very versatile and benefits from relatively low complexity, some important repercussions apply when used in GSM applications:

1. In order not to desensitize a nearby receiving mobile handset, a very stringent transmitted noise level of -79 dBm at the corresponding receive-band is specified, measured with a 100 kHz resolution bandwidth (RBW), as illustrated in Figure 2.9. This corresponds to an overall phase noise of -162 dBc/Hz at a worst-case offset frequency of 20 MHz at full power level of $+33$ dBm:

$$PN = -79 \text{ dBm} - 33 \text{ dBm} - 10 \log_{10}(100 \text{ kHz}) = -162 \text{ dBc/Hz}. \quad (2.4)$$

   Given that typical GSM power amplifiers (PAs) have a noise floor of -83 dBm to -84 dBm, the required phase noise at the RF IC output is calculated to be better than -165 dBc/Hz in the extreme case. In the case of a direct-conversion TX, in addition to the voltage controlled oscillator (VCO) and the local oscillator (LO) chain, the entire TX path such as the DACs, reconstruction filters, and the active or passive mixers [36] contribute to this noise. Due to such stringent requirements an external front-end filter and/or VCO may be inevitable, unless at the expense of increased power consumption.

2. Even though the GSM signal is only phase modulated, in the case of a direct-conversion architecture the linearity of the modulator and the PA is a concern. This may sound counter-intuitive at first, but can be understood by noting that in the switching mixers used in most modulators [36], a replica of the modulated input of opposite phase exists at the local oscillator (LO) third harmonic. When mixed down due to the third-order nonlinearity of the PA or the PA driver, this signal replica could cause degradation of the modulation spectrum. Resolving this issue could lead to a further increase in the TX power dissipation and/or complexity.

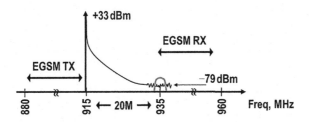

FIGURE 2.9

GSM TX noise in RX band.

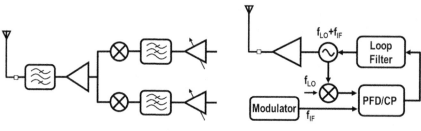

FIGURE 2.10

Direct-conversion TX vs. translational loop.

3. Despite operating the TX VCO at two or four times the output frequency in most common radios, a direct-conversion scheme suffers from the potential pulling caused by the PA running at 2 W, and drawing more than 1 A of current from the battery. This is certainly more prominent in the case of EDGE and 3G, where the signal is amplitude modulated as well.
4. Direct-conversion transmitters generally require good I-Q balance in the signal and LO path. Achieving that typically requires some kind of calibration which may add to the complexity. Moreover, the DC offset of the baseband section including the DAC and the following reconstruction filters needs to be corrected. As will be discussed in section VI, this issue could be more severe in the case of a 3G TX where a very wide gain range is required.

For these reasons GSM transmitters have traditionally adopted translational loops to eliminate the external front-end filter at reasonably low-power consumptions [7–9], as shown in Figure 2.10.

The PLL forces the VCO output to be a replica of the modulated input spectrum fed to the phase frequency detector (PFD), thus translating the intermediate frequency (IF) spectrum to radio frequency (RF). Unlike the direct-conversion case, here only the VCO and LO chain contribute to the far-out noise, as usually the PLL bandwidth is narrow enough to suppress the noise of the rest of the loop. Moreover, as the VCO and PA output have the exact same phase content, this type of transmitter is more immune to pulling.

The translational loop can be simplified to a direct-modulated PLL [10–13], where modulation is injected via a $\Delta\Sigma$ modulator by changing the fractional divider modulus over time (Figure 2.11). The main advantage is the elimination of the baseband IQ modulator needed in the translational loop, which could be a major noise and nonlinearity contributor. The LO needed for the mixing in the loop can be eliminated as well, which reduces the area and power consumption further. This typically requires the PLL to run off the available 26 MHz crystal oscillator, although one may choose to adopt an auxiliary PLL to ease some of the typical problems of the fractional PLLs caused by low-frequency reference signals [10,23]. Overall, if designed properly, the direct-modulated PLL can lead to substantial area and power savings.

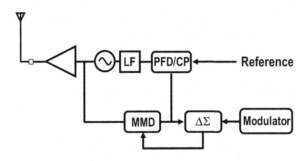

FIGURE 2.11

PLL-based GSM transmitter.

However, to achieve satisfactory performance several concerns must be addressed [26].

The main disadvantage of the PLL-based transmitter is its lack of support for amplitude-modulated standards as only phase modulation is created through the VCO. A time-varying AM signal component, such as that needed for EDGE or WCDMA modulation, can be produced by employing another $\Delta\Sigma$ modulator as shown in Figure 2.12 [27–30].

This modulates the PA driver, which can be thought of as a single-balanced mixer. Although an *open-loop small-signal polar* architecture as described here further adds to the transmitter complexity, all the disadvantages of the alternative direct-conversion approach discussed earlier hold here as well. Moreover, the PA driver, one of the biggest contributors to the TX power consumption, can operate closer to saturation compared to a linear TX system, further improving the power consumption. The polar TX also enjoys a simpler 50%, single-phase LO path, as opposed to the linear transmitters that typically require quadrature 25% LO, exacerbating the power efficiency

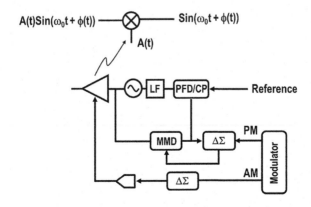

FIGURE 2.12

Polar transmitter for EDGE.

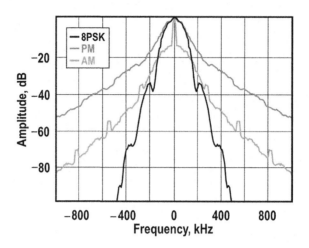

FIGURE 2.13

PM and AM spectrums for EDGE.

problem. Additionally, the polar architecture is "*backward compatible*" to that needed for GSM transmission and thus saves area. A compromise solution, consisting of a direct-modulated PLL for GSM mode of operation and a linear TX for EDGE or WCDMA mode of operation, is common but leads to higher power and cost.

Despite the power and area advantages of the polar transmitter, there are many challenges and design concerns that must be taken into account which will be briefly discussed here. A more detailed analysis is beyond the scope of this book, but can be found in [26].

1. Even though the EDGE signal is only 200 kHz wide, when broken into phase and amplitude domain, both the PM and AM spectrums grow considerably (Figure 2.13). Particularly the PM signal is very wide, only dropping by about 30 dB at 400 kHz offset, compared to 68 dB for the ideal EDGE spectrum. This demands substantially larger swings at the VCO input, thus raising the sensitivity to nonlinearities. Particularly if the PLL bandwidth is around 200 kHz or less, it will cause the loop gain to drop at 400 kHz and beyond.
2. In addition to the non-idealities of the AM and PM paths individually, the alignment between the AM and PM signals is critical as shown in Figure 2.15. In order not to degrade the EDGE spectrum significantly, a delay error of less than 20 ns is needed. Thus, the PLL bandwidth as well as that of the AM reconstruction filter must be tightly controlled.

2.1.2.2 Phase error and EVM requirements

Since the PLL loop BW is relatively narrow, the frequency-modulated (FM) equivalent GSM spectrum will be subject to distortion. Such distortion can be compensated for a digital pre-distortion block whose frequency response is the inverse of that

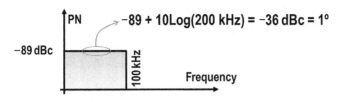

FIGURE 2.14

Phase error calculations for GSM.

of the PLL, hence providing an overall all-pass shape for the injected modulation. However, as the PLL characteristics tend to vary throughout the process, the overall response will generally not be flat, causing phase error degradation in the transmitted signal. This is potentially the biggest source of phase error degradation as shown in [26]. The unmodulated RMS phase error of the PLL arising from the in-band noise of the PLL sets the baseline and adds further to the phase error caused by BW variation. For example, an in-band phase noise of -89 dBc/Hz (-36 dBc or 0.019 across 200 kHz bandwidth) leads to an additional RMS phase error of 0.016 Rad or 1° as shown in Figure 2.14.

Similar to phase error for GSM modulation, PLL BW mismatch is the biggest contributor on EDGE EVM. As with GMSK mode, the unmodulated phase error sets the floor. For example, -89 dBc across 200 kHz leads to -36 dBc $= 10 \log(0.016)$, which is 1° RMS or 1.6% of EVM.

In the case of using a direct-conversion transmitter for the EDGE mode, the mask and EVM contributors will be similar to the ones described for 3G TX in the next section, where they will be discussed in detail.

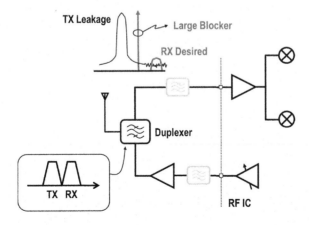

FIGURE 2.15

Full duplex challenges of 3G systems.

2.2 3/4G transceiver architecture design

The 3G standard as introduced initially in 2001 [4] is a spread-spectrum wideband CDMA system using a QPSK modulation scheme and supports up to 19 bands, the most widely used ones are band I (2110–2170 MHz for downlink, 1920–1980 MHz for uplink) and bands II, III, VIII, and V which are identical to the four GSM/EDGE bands shown in Table 2.1 (PCS, DCS, GSM 900, and GSM 850, respectively). Several other bands such as bands IV, VI, or IX are a subset of the aforementioned five bands, while the less commonly used band VII uses a more challenging higher frequency (2620–2690 MHz for downlink, 2500–2570 MHz for uplink). The channel spacing is 5 MHz, while the signal itself is about ± 1.92 MHz wide. The TX-RX separation ranges from 30 MHz for lowbands (although most commonly used lowbands are 45 MHz) and up to 190 MHz for band I. The recent drive for higher data rates has, however, introduced several additions to the original 3G, such as HSPA, which uses 16-QAM in RX mode, and HSPA+, which supports 64-QAM, and can achieve data rates of up to 21 Mbps.

Unlike GSM/EDGE, 3G is an FDD (frequency division duplexing) system. The *full-duplex* nature of the system, that is, the simultaneous operation of the RX and TX causes several challenges unique to 3G, although the spread spectrum nature of it [19–22] relaxes some of the blocking requirements. Ideally, an *external duplexer* realized by two *highly selective filters* separates the receive and transmit signals (Figure 2.15), but in practice, due to the finite isolation of the duplexer, some of the strong TX signals leak to the RX input, causing two issues: First, the TX noise falling in the RX band effectively degrades the receive noise figure. Second, when mixed by a large out-of-band blocker (for example, the blocker at *half-duplex frequency*) due to the front-end third-order nonlinearity, the RX is desensitized. To overcome these issues, external filters are traditionally placed at the TX and RX ports to suppress the TX noise and leakage, thus relaxing the *phase noise* and *linearity* requirements of the transceiver. In the case of the transmitter output, the SAW filter relaxes the noise requirement of the TX chain by providing some filtering, whereas the receiver SAW filter attenuates the TX residual leakage and any other blocker, thereby relaxing the linearity requirements of the RX chain. However, most modern 3G designs are SAW-less, which demands more stringent phase noise (for both RX and TX) and linearity (for RX). In this section we will look into some of these requirements in more detail.

2.2.1 3G receiver requirements

Most recent 3G receivers usually adopt a zero-IF architecture shown in Figure 2.2 [19–22]. As the received signal is wide compared to 2G, common issues such as $1/f$ noise and DC offset are less problematic, thus justifying a zero-IF design due to its simplicity and low power consumption. On the other hand, the wideband 3G signal would require a large IF (a few MHz) if one were to use a low-IF scheme, which translates to higher power consumption for the ADC and IF analog blocks, without

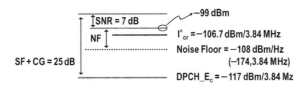

FIGURE 2.16

3G RX noise figure.

gaining much performance. The receiver is SAW-less; however, it benefits from the filtering provided by the duplexer, which is similar to a 2G SAW filter for a given band. Therefore the out-of-band blockers (except for the TX leakage and related blockers) are generally not a major concern.

2.2.1.1 Sensitivity

The receiver noise figure is calculated based on a reference sensitivity requirement of $-117\,\mathrm{dBm}/3.84\,\mathrm{MHz}$ (for band I as an example), and for a 12.2 kbps reference measurement, with a spreading factor of 128 (or 21 dB) as shown in Figure 2.16.

The spreading factor (SF) is the ratio of the chip rate, 3.84 MHz, to the actual data rate, which is 30 kbps in this measurement. In general, depending on the physical conditions of the link and the quality of the reception, the spreading factor ranges from 4 to 512, where the SF of 4 corresponds to the highest throughput but requires a stronger signal. Since the QPSK requires a modem SNR of 7 dB, with 4 dB of coding gain and 21 dB of spreading factor, the effective SNR is $7\,\mathrm{dB} - 4\,\mathrm{dB} - 21\,\mathrm{dB} = -18\,\mathrm{dB}$. Thus the NF is $174 - 117 + 18 - 10 \times \log(3.84\,\mathrm{MHz}) = 9.2\,\mathrm{dB}$. Unlike the 2G receiver where the sensitivity is merely set by the receiver thermal noise, the NF here is derived by several other factors as well, such as the TX noise in the RX band, the receiver second-order non linearity, and the reciprocal mixing. We will discuss the impact of TX noise in the next section (Eq. (2.5)). The receiver second-order non-linearity effectively amplitude de-modulates the TX leakage, which results in a wide signal spread at twice the desirable signal bandwidth at DC, as shown in Figure 2.17.

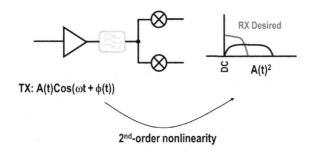

FIGURE 2.17

TX leakage causing second-order nonlinearity concerns.

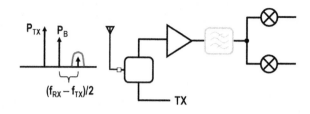

FIGURE 2.18

Out-of-band IIP3 of a 3G receiver.

Similar to our IIP2 analysis for the 2G receivers, the required IIP2 for the 3G RX can be calculated for a given TX leakage. Assuming a duplexer isolation of 45 dB in the transmit frequency, and 4 dB loss for the duplexer/switch, the TX leakage is $24 + 4 - 45 = -17$ dBm at the LNA input, assuming a full transmit power of 24 dBm at the antenna. As $IIP2 = 2 \times P_B - Sensitivity + SNR$, with $P_B = -17$ dBm, an IIP2 of $(2 \times -17 - 13) + 117 - 18 = +52$ dBm will cause a sensitivity of -117 dBm. The 13 dB is a constant found through simulations to consider the spreading on the TX envelope squared (Figure 2.18). Depending on the duplexer, most 3G receivers target an IIP2 of better than $+45$ dBm specified at the duplex frequency (190 MHz away for band I for instance). With a SAW filter placed after LNA however, the TX leakage is attenuated substantially and the IIP2 is relaxed accordingly. Similarly a phase noise of PN at the duplex frequency on the RX VCO/LO defines a thermal noise figure of $174 + P_{TX} + PN$, where P_{TX} is the TX leakage. Now a phase noise of -155 dBc/Hz at the duplexer frequency alone results in a noise figure of 2 dB, which adds to the overall budget. Unlike the case of 2G, where the sensitivity is relaxed by 3 dB in the presence of the blockers, here the mentioned impacts all add up in the extreme case.

Now assuming 3 dB insertion loss for the duplexer, 1 dB loss for the switch, and another 1 dB degradation due to the impacts of second-order nonlinearity, TX noise, and reciprocal mixing, then the thermal receiver noise figure is 4.2 dB at worst case. With 1.5 dB margin for PVT, the receiver typical thermal noise figure is about 2.5 dB, very similar to the GSM. Note that one may trade the receiver thermal noise for a more stringent IIP2 or phase noise, for example, or perhaps a better (more expensive) duplexer with higher isolation.

The impact of blockers is generally less important than GSM, except for the cases driven by the TX leakage such as duplex IIP2, which was discussed previously. Another example is shown in Figure 2.18 as follows.

A blocker at half-duplex frequency mixed with the TX leakage results in an IM3 signal down-converted on top of the desirable signal at baseband due to the third-order nonlinearity of RX front-end. The sensitivity is allowed to be relaxed by 3 dB in the out-of-band blocker scenarios, and the IIP3 is calculated as follows: $2 \times IIP3 = P_{TX} + 2 \times P_B - Sensitivity + SNR$. The TX leakage was found to be -17 dBm for 45 dB of isolation, and assuming 30 dB of filtering due to the duplexer and -15 dBm blocker power as specified by 3GPP standard, IIP3 is found to be

$(-17-2\times45+114-18)/2=-5.5\,$dBm. With PVT and production margins, and some room left for other factors such as phase noise or second-order nonlinearity as discussed before, the half-duplex IIP3 is typically specified to be close to 0 dBm, which is much more challenging than what is needed for GSM/EDGE ($-13\,$dBm at 3 MHz). Again a filter placed between the LNA and mixers would help at the expense of size and cost. Alternatively one may choose either integrated LC notch filtering [21], or use high-Q n-path filtering [14–15] shown in Figure 2.20 [17]. In the front-end of Figure 2.19, the LNA is replaced by a low-noise transconductor, driving current-mode passive mixers. Due to reciprocity, the passive mixers up-convert the low-pass baseband impedance to RF providing high-Q selectivity on the LNA output, thus suppressing the TX leakage. With this prototype IIP3 of better than 0 dBm with 2.5 dB noise figure is achieved.

The maximum input level is $-25\,$dBm for it is typically easily met through gain control in the receiver.

2.2.1.2 Blocking requirements

3GPP specifies an adjacent channel (at +5 or $-5\,$MHz away) selectivity (ACS) of 33 dB specified for two cases of desired signal at 14 or 41 dB above the reference sensitivity. Since desirable signal is relatively strong, typically noise is not a factor and,

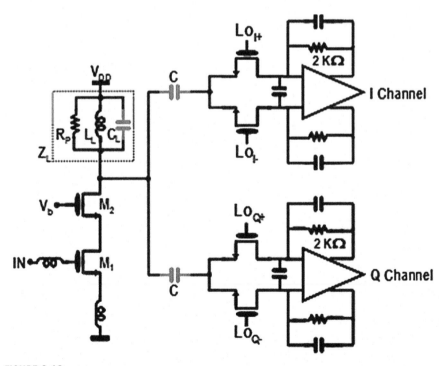

FIGURE 2.19

An example of a 3G RX front-end.

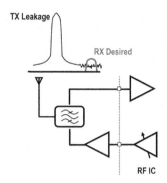

FIGURE 2.20

3G TX noise in RX band.

similar to GSM adjacent blockers (Figure 2.5), the main requirements are imposed on the baseband filtering (order and stop band rejection) as well as the ADC. If for cost (silicon area) reasons a low-order filter (for example a third-order filter) is used, since the filter needs to be about 2 MHz wide, the rejection at 5 MHz away is relatively small, which puts more burden on the ADC.

In the case of in-band blocking, the signal is specified to be 3 dB above the reference sensitivity (for band I, for example, then the signal power is about −104 dBm), whereas the in-band blockers at ±10 MHz away have a mean modulated power of −56 dBm, and at ±15 MHz or above it is −44 dBm. Thus at 10 MHz, the blocker is 48 dB stronger, for example. As with ACS, this sets the trade-off between the low-pass filter order and the ADC dynamic range. The out-of-band blockers are set to be maximum at −15 dBm, which is far less than the GSM requirements. An example of out-of-band IIP3 was discussed previously.

Finally, a narrow-band GMSK modulated blocker at ±2.7 MHz away is specified to be at −57 dBm with the signal at 10 dB above the sensitivity ($\hat{I}_{OR} = -96.7$ dBm for band I, for example). Therefore the blocker is about 40 dB higher than the signal and is subject to almost no filtering as the channel-select filter is about 2 MHz wide. Leaving a 10 dB upper margin for AGC error, uncanceled DC offset, assuming the ADC noise floor needs to be 20 dB below the signal, this corresponds to a dynamic range of about 70 dB for the ADC.

The inter-modulation blockers are specified at −46 dBm and at ±10 MHz and ±20 MHz away from the signal at 3 dB above the reference sensitivity. Assuming a typical target of −40 dBm for the RF IC to leave some margin for PVT and other production concerns, since IIP3 = $3P_B$ − Sensitivity + SNR, the required IIP3 is calculated to be −12 dBm. The actual SNR needs to be higher than the worst case −18 dB used before to leave some room for other factors including the RX thermal noise and TX leakage related issues, thus requiring an IIP3 of usually better than the −10 dBm met by the radio.

For HSDPA and other subsets of 3G similar requirements are specified, and usually meeting the basic one for original 3G is sufficient, with the exception of the SNR, which is higher for higher data rate modulations such as HSDPA and HSPA+.

2.2.2 3G transmitter requirements

The advantages of polar transmitters described in Section 2.1 are not limited to GSM/EDGE applications. In fact, arguably wider band applications with more complex modulation schemes (and thus higher peak-to-average ratio) such as 3G may potentially benefit more. Lower power and area due to a simpler LO path and more efficient (*less linear*) PAD, and concerns such as TX noise in receive-band, are primary motivating factors. Even though the *duplexer* provides some filtering (unlike 2G where the PA directly connects to the antenna), the far-out noise requirements are not much relaxed (Figure 2.20). For example with a duplexer isolation of 48 dB in the receive-band, the receiver noise floor due to the transmitter far-out phase noise of −160 dBc/Hz alone is:

$$\text{Noise Floor} = -160 \text{ dBc} - 48 \text{ dB} + (24 \text{ dBm} + 4 \text{ dB}) = -180 \text{ dBm/Hz}. \quad (2.5)$$

Here, a 4 dB duplexer/switch loss and the full power of 24 dBm at the antenna have been assumed. If the receiver thermal noise figure is now 3 dB for instance (corresponding to a noise floor of −171 dBm/Hz), the TX phase alone could cause over 0.5 dB degradation in the composite noise figure and thus the receiver sensitivity. This kind of phase noise is particularly challenging to achieve in a *linear transmitter* where the baseband path and the modulator contribute substantially.

Despite the disadvantages mentioned above, due to the challenges of polar implementation so far all mainstream 3G transmitters are of the direct up-conversion type [20–22,31,32]. Finding a reliable method of wideband phase modulation and suppressing unequal delay in the two separate AM and PM paths are the main issues as discussed before. Whereas EDGE requires a PM bandwidth of about 800 kHz and can tolerate AM-PM delay mismatch on the order of tens of nanoseconds, the WCDMA PM signal requires a bandwidth of at least 8 MHz as shown. Moreover, simulations indicate that a path delay mismatch of less than 2 ns is required to ensure an acceptable adjacent channel leakage ratio (ACLR), about 10 times more stringent delay mismatch than is needed in EDGE mode. Therefore we will mainly focus on linear 3G transmitters in this section. A detailed system analysis of polar 3G transmitters can be found in [24], and an example of digital PLL-based 3G polar transmitter is described in [25].

2.2.2.1 Output power and gain control

According to 3GPP, all the bands are required to transmit a maximum output power of +24 dBm at the antenna with +1/−3 dB tolerance for operation in power class 3. The minimum power required is −50 dBm. Since a 3G PA is linear, most of the gain control is performed in the RF IC while PAs provide 10–20 dB of gain control to improve the efficiency at lower powers. With a 3 dB margin for each end, and assuming a PA gain control of 10 dBc, the RF IC transmitter then needs to achieve over 70 dB of gain control. The PA has a linear maximum gain of about 25–30 dB, and with 4 dB loss of the duplexer/switch, the transmitter should put out about 0 dBm in typical conditions and less than −70 dBm of minimum power. The power control is performed in steps of 1, 2, or 3 dB with a corresponding tolerance of ±0.5 dB, ±1 dB, and ±1.5 dB. This tight power resolution along with the wide range of power control is among the most challenging characteristics of a 3G transmitter. Typically at higher power levels a closed loop power control is implemented to meet the requirements along with several steps of factory calibration. This will be discussed more in the "Conclusions" section.

2.2.2.2 EVM and mask

The 3GPP standard requires an EVM of better than 17.5% for the overall TX across a wide gain range of 74 dB. In an IQ 3G transmitter there are several contributors to EVM such as LO feed-through and IQ imbalance. For example, an LO feed-through and an IQ imbalance of −40 dBc, along with −40 dBc integrated phase noise for the TX VCO, lead to:

$$\text{EVM} \approx \sqrt{10^{\frac{\text{LOFT}}{10}} + 10^{\frac{\text{IQ}}{10}} + 10^{\frac{\text{PN}}{10}}} = \sqrt{3} = 1.7\%. \tag{2.6}$$

Most advanced radios target less than 3% EVM at maximum power. The common elaborate calibration schemes must be exploited, particularly at low powers where the feed-through created by the mismatches in the modulator may not even be detectable at the RF IC output. Moreover, the baseband IQ low-pass filters ripple and non-flat group delay adds to the EVM. Exact similar factors limit the EDGE EVM if one chooses to use an IQ linear transmitter as opposed to polar, which was discussed in the previous section.

The 3G mask is defined by adjacent channel leakage ratio (ACLR). The most challenging one is the ACLR1 at 5 MHz away, which is specified to be less than −33 dBc measured at 3.84 MHz bandwidth. ACLR2 at 10 MHz needs to be lower than −43 dBc;

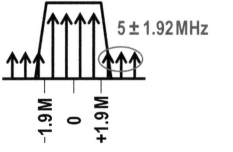
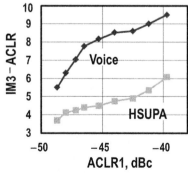

FIGURE 2.21

3G TX ACLR calculations.

this is usually automatically met if the ACLR1 is satisfied. Unlike the GSM transmitter, phase noise has a lesser impact on the in-band phase noise, whereas the linearity of the IQ modulator and PA driver are key contributors. Shown in Figure 2.21, the WCDMA signal can be thought of as several close-by tones, each of which creates IM3 sidebands due to the TX third-order nonlinearity that falls outside the desirable signal spectrum. Therefore, to a first-order approximation, the TX IIP3 is directly related to ACLR and spectrum re-growth. A more detailed analysis based on ADS simulations, shown below, can be used to estimate the TX linearity from a two-tone test for a given ACLR.

Note that the nonlinearity and the portion of the IM3 components falling in band will affect the EVM too. However, meeting the ACLR will typically make the IM3 a non-major EVM contributor. The phase noise at 5 MHz is a factor; however, meeting the RX-band and in-band phase noise usually suffices.

As a 3G transmitter needs to achieve an ACLR1 of better than -33 dBc at the antenna, assuming a worst-case scenario ACLR of -37 dBc for the PA (which are optimized for efficiency and thus sacrifice some linearity), and 2 dB of production margin, the transmitter contribution is set to -40 dBc or less. From Figure 2.18 then the IM3 needs to be lower than $-40+9=-31$ dBc. At a maximum radio output power of 3 dBm, the OIP3 is calculated to be 15.5 dBm, as shown in Figure 2.22. The output 1 dB compression is then 5.5 dBm.

As for the impact of phase noise, assuming a relaxed phase noise of -130 dBc at 5 MHz (for 2G we need -120 dBc at 400 kHz, or roughly better than -140 dBc at 5 MHz, assuming 20 dB/Dec roll-off), then according to Eq. (2.3), the ACLR1 due to this phase noise will be $-130+10\times\log(3.84\,\text{MHz})=-64$ dBc, while the RF IC requirement is -40 dBc. Even though this is an approximation, it does show that the impact of phase noise is less dominant in a linear 3G transmitter.

2.2.3 **LTE transceiver requirements**

The Long-Term Evolution (LTE) supports up to 300 Mbps data rate for downlink and up to 75 Mbps for uplink, with the ability to manage fast-moving mobile objects. The modulation is OFM based and supports a flexible bandwidth of 1.4–20 MHz (vs.

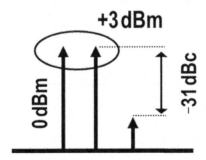

FIGURE 2.22

3G TX linearity.

5 MHz in 3G). Similar concerns mentioned for the case of an FDD 3G transceiver exist here as well. For the LTE standard, the out-of-band filtering requirements are the same as those required for 3G (−15 dBm, which is the worst case for LTE, thus still the dominant requirement is for GSM). Due to the wider bandwidth, however, the in-band blocking requirement for LTE is more stringent. This requirement, however, mainly imposes greater challenges on the integrated channel-select filter design, and less on the RF front end. Note that, similar to 3G, LTE must support the FDD option as well, and the issues with TX leaking to the receiver and causing stringent IIP2, out-of-band IIP3, and phase noise stay more or less the same. The maximum output power is +23 dBm (with ±2 dB of tolerance) which is 1 dB less than 3G. The minimum output power is −40 dBm for all the bandwidths, so the overall gain control range is 11 dB less than 3G.

The noise figure and IIP3 are roughly the same as 3G, while the IIP2 is about 7 dB more stringent (+52 dBm or better) as the 13 dB factor for the 3G TX squared spreading (Figure 2.18) is about 6 dB in the case of LTE. Similarly it can be shown that the transmitter requirements as far as linearity or phase noise is concerned are similar to the case of 3G. For example, for the highest throughput case of 20 MHz bandwidth and 64QAM, the peak-to-average ratio is about 6 dB (3 dB in the case of 3G), while the transmit output power is 1 dB less.

One distinct feature of LTE is the receiver diversity, that is two identical receivers running from the same VCO/LO but connected to two separate antennas. As shown in Figure 2.23, this creates two distinct advantages:

1. High-rate legacy data can be received more robustly by optimally combining the multiple chains' signals (*diversity combining*).
2. Higher-rate data can be received by having the transmitter send independent data on multiple antennas (*spatial multiplexing*).

For example as shown in Figure 2.23 signals received on multiple antennas can be combined to create a more benign composite channel, and thus combined signal energy increases SNR.

To support further higher data rates, another feature added to LTE (known as LTE-Advanced) is carrier aggregation (CA) shown in Figure 2.24.

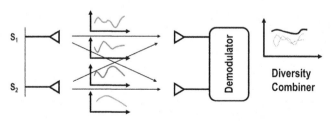

FIGURE 2.23

Diversity in LTE receivers.

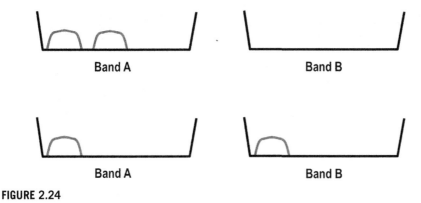

FIGURE 2.24

Carrier aggregation in 4G systems.

The LTE Advanced targets data throughput in the downlink of up to 1 Gbps. To achieve the higher data rates the overall bandwidth is increased by allocating more than one 20 MHz channel for use. This could be intra-band, that is, the two channels are within the same band of operation (whether adjacent or not), or it could be two different channels in two different bands of operation (inter-band) as shown in Figure 2.21. Apart from complexity and the need for two separate VCO/PLLs, the pulling between the two VCOs is a potential concern. From this perspective, the inter-band mode is advantageous, as the two channels are far enough away in frequency.

2.3 Handset calibration

Due to the stringent requirements imposed on cellular transceivers, there are typically a number of calibrations needed to enhance the performance and compensate for the shortcomings of the modern CMOS radios. There are two categories, factory calibrations and automatic internal calibrations. We will discuss them briefly here and provide a few examples. The exact nature of the calibration certainly depends on the specific radio architecture and circuit design while there are some that are common among most modern radios today.

2.3.1 Factory calibrations

The factory calibration is performed only once during the handset lifetime, and is desirable to be minimized due to cost reasons. Obviously temperature and aging-related variations are not covered by the factory calibration. Below are some of the most common ones:

1. *Crystal oscillator tuning:* This is performed through the coarse crystal adjustment as was discussed in the previous section. Depending on the coarse resolution and the calibration accuracy, this brings the crystal frequency within a few ppm, and the residual error is handled by the AFC (automatic frequency calibration) [35].

2. *Receiver gain calibration:* The exact gain of the receiver is important for several reasons. First an RSSI (received signal strength indicator) of better than 2 dB is needed, while the receiver gain can vary substantially more than that over PVT and extreme conditions. Secondly, to utilize the full dynamic range of the ADC, the receiver gain must be well known.

3. *Transmitter gain calibration:* The output power of the system at the antenna must be very accurate depending on the output power level at a given gain. Both the transceiver and the PA gain vary and to accommodate for that a closed loop power control (CLPC) scheme is usually used, which relies on the peak detector available in most linear PAs. Despite that, the total gain including that of the PA is calibrated one time at the factory to provide an accurate absolute reference point. It is desirable to only perform this at one gain settings to avoid extra cost. Moreover, the slope and characteristics of the peak detector may also need to be corrected to ensure a good CLPC accuracy.

2.3.2 Automatic calibrations

This includes both a one-time calibration at the startup as well as continuous adjustments over bursts (which tracks temperature or any other transient variations). The main intention is to overcome the common issues of low-cost CMOS, such as poor matching, DC offsets, and second- or third-order nonlinearity. All these calibrations are automatic within the radio, and do not require external measurement (unlike the factory calibration) and therefore are typically free of any cost. However, they may increase the handset boot time or may have extra power consumption or area implications. In this section we will discuss some of the most common ones as examples.

One of the most routine automatic calibrations common in almost every EDGE/3G linear transmitter is the correction for LO feed though (LOFT) and IQ imbalances that can adversely affect the EVM. A typical correction method, shown in Figure 2.25, involves sending a test tone through the TX DAC and monitoring the TX output envelope through an integrated peak detector.

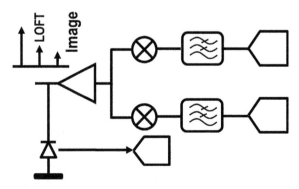

FIGURE 2.25

TX LOFT and IQ calibration.

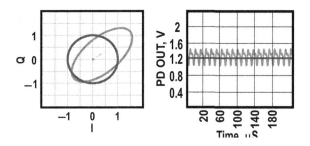

FIGURE 2.26

TX constellation for non-ideal LOFT and IQ imbalance.

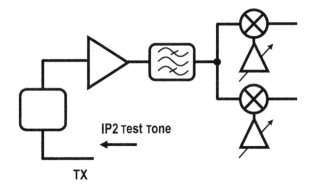

FIGURE 2.27

An example of a 3G receiver IIP2 enhancement loop.

In the absence of any mismatch or DC offset, the output constellation will be a perfect circle at origin whereas the DC offsets will move the circle from the origin, while IQ imbalances will distort the circle as shown in Figure 2.26.

For the receiver, there are a number of common calibrations used such as DC offset correction, IQ calibration, and IIP2 correction. Given the very stringent IIP3 requirement of 3G and LTE receivers, it is common to enhance the RX IIP2 by adopting an automated second-order nonlinearity correction as shown in Figure 2.27.

CONCLUSIONS

Cellular transceivers must satisfy very stringent requirements. This includes both the ability to achieve low noise for RX and high output power for TX along with good quality (SNR or EVM), while having a very high blocking tolerance or stringent mask requirements, which makes them unique yet much more difficult to analyze compared to other applications, such as WLAN or WPAN. In this chapter we offered a system level analysis of cellular RF transceivers for 2/3/4G applications. Details of RF cellular standard were presented, followed by a discussion of various radio architectures. Practical circuit level requirements of the radio suitable for today's CMOS processes were derived with discussions on various architectural ideas.

References

[1] ABI Research, Mobile Device Semiconductors Market Data 2Q2010 Report. <www.abiresearch.com>.

[2] ABI Research, Mobile Devices Annual Market Overview 4Q2009 Report. <www.abiresearch.com>.

[3] 3GPP TS 05.10 V8.12.0 (2003–08), Digital Cellular Telecommunications System (Phase 2+); Radio Subsystem Synchronization (Release 1999).

[4] 3GPP TS 25.101 V6.6.0 (2004–12), User Equipment (UE) Radio Transmission and Reception (FDD) (Release 6).

[5] 3GPP TS 26.101 V6.6.0 (2004–12), User Equipment (UE) Radio Transmission and Reception (FDD) (Release 10).

[6] A. A. Abidi, Direct-conversion radio transceivers for digital communications, *IEEE Journal of Solid-State Circuits*, vol. 30, no. 12, pp. 1399–1410, 1995.

[7] O. E. Erdogan et al., A single-chip quad-band GSM/GPRS transceiver in 0.18 μm standard CMOS, *ISSCC Digest of Technical Papers*, pp. 318–319, February 2005.

[8] O. Bonnaud et al., A fully integrated SoC for GSM/GPRS in 0.13 μm CMOS, *ISSCC Digest of Technical Papers*, pp. 482–483, February 2006.

[9] R. Pullela et al., An integrated closed-loop polar transmitter with saturation prevention and low-IF receiver for quad-band GPRS/EDGE, *ISSCC Digest of Technical Papers*, pp. 112–113, February 2009.

[10] H. Darabi et al., A fully integrated quad-band GPRS/EDGE radio in 0.13 μm CMOS, *ISSCC Digest of Technical Papers*, pp. 206–207, February 2008.

[11] J. Mehta et al., A 0.8 mm^2 all-digital SAW-less polar transmitter in 65 nm EDGE SoC, *ISSCC Digest of Technical Papers*, pp. 58–59, February 2010.

[12] G. Puma et al., Integration of a SiP for GSM/EDGE in CMOS technology, *ISSCC Digest of Technical Papers*, pp. 210–211, February 2008.

[13] M. Elliott et al., A polar modulator transmitter for EDGE, *ISSCC Digest of Technical Papers*, pp. 188–189, February 2004.

[14] L. Franks and I. Sandberg, An alternative approach to the realizations of network functions: N-path filter, *Bell System Technical Journal*, pp. 1321–1350, 1960.

[15] D. V. Grunigen et al., An integrated CMOS switched-capacitor bandpass filter based on N-path and frequency-sampling principles, *IEEE Journal of Solid-State Circuits*, vol. SC-18, no. 6, pp. 753–761, 1983.

[16] A. Mirzaei et al., Analysis and optimization of direct-conversion receivers with 25% duty-cycle current-driven passive mixers, *IEEE Transactions on Circuits and Systems Part I*, vol. 47, no. 9, pp. 23530–23566, 2010.

[17] A. Mirzaei, J. Leete, X. Chen, and H. Darabi, A frequency translation technique for SAW-less 3G receivers, in *VLSI Symposium*, June 2009.

[18] A. Mirzaei et al., Analysis and optimization of current-driven passive mixers in narrowband direct-conversion receivers, *IEEE Journal of Solid-State Circuits*, vol. 10, pp. 2678–2688, October 2009.

[19] D. Kaczman et al., A single-chip 10-band WCDMA/HSDPA 4-band GSM/EDGE SAW-less CMOS receiver with DigRF 3G interface and 90 dBm IIP3, *IEEE Journal of Solid-State Circuits*, vol. 3, pp. 718–739, 2009.

[20] B. Tenbroek et al., Single-chip tri-band WCDMA/HSDPA transceiver without external SAW filters and with integrated TX power control, *ISSCC Digest of Technical Papers*, pp. 202–203, February 2008.

[21] C. Jones et al., Direct-conversion WCDMA transmitter with −163 dBc/Hz noise at 190 MHz offset, *ISSCC Digest of Technical Papers*, pp. 336–337, February 2007.

[22] Q. Huang et al., A tri-band SAW-less WCDMA/HSPA RF CMOS transceiver with on-chip DC-DC converter connectable to battery, *ISSCC Digest of Technical Papers*, pp. 60–61, February 2010.

[23] H. Darabi et al., A Quad-Band GSM/GPRS/EDGE SoC in 65 nm CMOS, *IEEE Journal of Solid-State Circuits*, vol. 4, April 2011.

[24] M. Youssef, A. Zolfaghari, H. Darabi, and A. Abidi, A low-power wideband polar transmitter for 3G applications, *ISSCC Digest of Technical Papers*, February 2011.

[25] Z. Boos et al., A fully digital multi-mode polar transmitter employing 17b RF DAC in 3G Mode, *ISSCC Digest of Technical Papers*, February 2011.

[26] H. Darabi, H. Jensen, and A. Zolfaghari, Analysis and design of small signal polar transmitters for cellular applications, *IEEE Journal of Solid-State Circuits*, no. 6, June 2011.

[27] Y. Akamine et al., A polar loop transmitter with digital interface including a loop-bandwidth calibration system, *ISSCC Digest of Technical Papers*, pp. 348–349, February 2007.

[28] R. B. Staszewski et al., A 24 mm^2 quad-band single-chip GSM radio with transmitter calibration in 90 nm Digital CMOS, *ISSCC Digest of Technical Papers*, pp. 208–209, February 2008.

[29] A. Hietala, A quad-band 8PSK/GMSK polar transceiver, *IEEE Journal of Solid-State Circuits*, vol. 5, pp. 1131–1141, 2006.

[30] T. Sowlati et al., Quad-band GSM/GPRS/EDGE polar loop transmitter, *IEEE Journal of Solid State Circuits*, vol. 12, pp. 2179–2189, 2004.

[31] M. Cassia et al., A low-power CMOS SAW-less quad band WCDMA/HSPA/HSPA+/1X/EGPRS transmitter, *IEEE Journal of Solid-State Circuits*, vol. 44, no. 7, pp. 1897–1906, July 2009.

[32] T. Sowlati et al., Single-chip multi-band WCDMA/HSDPA/HSUPA/EGPRS transceiver with diversity receiver and 3G DigRF interface without SAW filters in transmitter/3G Receiver, *ISSCC Digest of Technical Papers*, pp. 116–117, February 2009.

[33] A. Mirzaei et al., A 65 nm CMOS quad-band SAW-less receiver for GSM/GPRS/EDGE, *IEEE Journal of Solid-State Circuits*, vol. 4, April 2011.

[34] I. Lu et al., A SAW-less GSM/GPRS/edge receiver embedded in a 65 nm SoC, *ISSCC Digest of Technical Papers*, pp. 364–365, February 2011.

[35] Y. Chang et al., A differential digitally controlled crystal oscillator with a 14-bit tuning resolution and sine wave outputs for cellular applications, *IEEE Journal of Solid-State Circuits*, vol. 2, February 2012.

[36] H. Darabi and A. A. Abidi, Noise in RF-CMOS mixers: a simple physical model, *IEEE Journal of Solid-State Circuits*, vol. 35, no. 1, pp. 15–25, 2000.

Low-Noise Amplifiers for Cellular Wireless Infrastructure

3

Domine Leenaerts

NXP Semiconductors, Eindhoven, The Netherlands

INTRODUCTION

In the mid-1990s research groups at various academic institutions started exploring CMOS technology to realize low-noise amplifiers (LNAs) operating at radio frequencies (RF). The preferred technology up to that moment to design in was a bipolar/BiCMOS process in order to get low-noise and high-linearity performance at RF. Almost two decades later, most commercial handheld terminals for mobile communication are based on a CMOS system-on-chip (SoC) approach, which includes the radio and at least parts of the baseband of the complete system functionality. This SoC includes typically more than one LNA to cope with multi-band, multi-mode cellular standards, e.g. GSM900, GSM1800, and WCDMA.

The fact that the LNA could be integrated in CMOS was due to two main reasons. First of all, the scaling trend in CMOS ensured that the MOS devices became fast enough to operate at RF cellular frequencies with good intrinsic noise performance. Secondly, the system specifications for cellular handheld terminals were defined such that the derived noise and linearity requirements for the receiver chain and LNA were difficult, but not impossible, to realize in modern CMOS technologies. The noise figure (NF) target for the cellular LNA is typically in the order of 2 dB with an associated 1 dB output compression point (OCP1dB) of approximately −5 dBm. Typically, a fully integrated LNA realizes an input return loss (IRL) of around 10 dB. Clearly, the fact that CMOS became the mainstream technology helped in the cost reduction of the overall handheld terminal and enabled faster ramp-up as a consumer product.

CMOS-based low-noise amplifiers with a better NF than 2 dB do exist. Examples can be found in GPS solutions and in astronomy applications; however, these solutions pay a price in achieved linearity and compression point.

On the other end of the spectrum we have the cellular wireless infrastructure, which is a professional market segment where performance rather than cost is the most important aspect. As we will see later, the derived performance requirements for the LNA are orders more difficult than for the handheld terminal. As a consequence, the majority of LNA circuits and modules in current macro and small cell base stations are realized in III–V compound technologies like GaAs. However, the trend in macro cell base stations is to move away from ground base stations towards

Manganaro: Advances in Analog and RF IC Design. http://dx.doi.org/10.1016/B978-0-12-398326-8.00003-0

tower-top-mounted base stations with, as a consequence, a reduced size and power budget. A move from III–V technologies to silicon-based technologies will enable this trend as silicon allows for denser integration and potentially lower power consumption. However, intrinsically a pHEMT GaAs device has superior noise and linearity performance over a silicon HBT device and MOS device. This will put severe design constraints on the design of a silicon-based LNA for wireless infrastructure.

3.1 LNA specifications

For a macro base station cell (or wide area base station) the 3GPP specifications specify a sensitivity level of -121 dBm for a data rate of 12.2 kbps in a bandwidth of 3.84 MHz [1]. This reveals a maximum total receiver noise figure at the antenna reference point (A.R.P.) of just 3 dB. Taking into account the losses of the duplexer filters between the A.R.P. and the input of the receiver and the noise due to the remainder of the receiver, the required NF of the first LNA must be below 1 dB. Even small cells (medium-range base stations and local-area base stations) require a sub-1 dB NF for the first LNA in the receive path.

The linearity requirements for the receiver are determined by the in-band inter-modulation characteristics rather than by the out-of-band inter-modulation requirements. The blocking requirements may be applied for the protection of base station receivers when for instance GSM900 or DCS1800 are co-located with the base station. With an interference power of $+16$ dBm and a duplexer isolation of $+30$ dB (in-band coupling loss, specified by 3GPP standard), the transmit signal at the receiver input is still -14 dBm. With a defined desensitization level of -109 dBm, that is 6 dB above sensitivity level, the input intercept point due to the third harmonic (IIP3) is -6 dBm for the complete receiver. Typically base station LNAs require an IIP3 around $+10$ dBm for maximum gain or beyond $+30$ dBm output IP3 (OIP3) for 20–25 dB power gain.

The transmitted power, received at the A.R.P, may not compress the receiver, leading to an input CP1dB (ICP1dB) of -13 dBm or better. Therefore the LNA needs to be designed to cope with this compression point level too. Typically base station LNAs have an ICP1dB around 0 dBm and assuming again 20 dB power gain, the output CP1dB (OCP1dB) is 20 dBm or better.

A very good input return loss (IRL) is important as the filter characteristic of the preceding duplexer filter is sensitive to the filter's load matching. Normally an IRL of better than 20 dB is required, i.e. the magnitude of s_{11} is lower than -20 dB. The output return loss (ORL) has a secondary importance and can be traded off for better IP3, but still needs to be better than 10–15 dB.

3.2 pHEMT-GaAs-based LNAs

In GaAs technology, the low-noise operation of a pHEMT device requires a bias current around 10–15% of the drain current at peak cut-off frequency (f_T) with a gate-source voltage equal to 0 V (assuming depletion-mode devices). However, high-linearity operation requires a much higher bias current, typically 70–80% of peak f_T. In a single-device LNA design a trade-off needs to be made between noise and linearity as only one bias current can be set. With a cascaded design the first and second stage can be biased independently and at different conditions, providing more design flexibility to find an optimal balance between noise and linearity.

3.2.1 Single-device LNA

In a single-device LNA design, noise and linearity requirements need to be met by optimal biasing and matching of the single pHEMT device. The intrinsic NF_{min} of a GaAs pHEMT device is approximately 0.4–0.6 dB at 2 GHz under optimal bias conditions. Moving away from this operating point towad a peak f_T bias point provides good linearity at the price of higher noise levels. But as the intrinsic noise level is low, an overall NF below or close to 1 dB is still possible.

A typical single device LNA is shown in Figure 3.1 (left part). The structure is known as a common-source (CS) device topology. A high-pass impedance-matching network consisting of L_1 and C_1 is used to match the input impedance to 50 Ω. A similar network, L_2 combined with C_2, is used to match the output impedance of the LNA to 50 Ω. Resistor R_3 provides low-frequency termination for the pHEMT device. The passive biasing is realized by the resistive network formed by R_1 and R_2. The voltage of this voltage divider is derived from the drain voltage of the pHEMT device. In this way a form of voltage feedback has been implemented which helps to keep the drain current constant.

Passive biasing cannot guarantee a constant quiescent bias point over temperature and process. Therefore many GaAs-based LNAs provide active biasing, for which an example is shown in Figure 3.1 (right part). In this case the bipolar PNP device Q_1 will ensure a constant drain current through the pHEMT device. The base voltage of Q_1 is set by R_1 and R_2. As the base-emitter voltage drop is technology

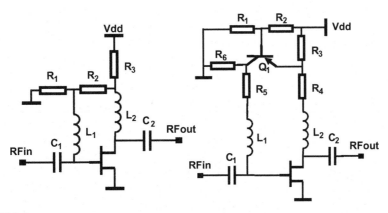

FIGURE 3.1

Single-pHEMT LNA with passive (left) and active (right) biasing.

Table 3.1 Overview of Several LNA Solutions on the Market, Targeting 1.95 GHz WCDMA

Performance @1950 MHz	Single-Stage LNA			Cascaded LNA	
	ADL5521	ATF-54143	RF3863	SKY65040-360LF	MGA-14516
Pdiss (mW)	300	180	450	325	775
Gain (dB)	16	15	15	25	32
NF (dB)	1.0	0.5	0.8	0.7	0.7
OIP3 (dBm)	35	36	35	35	38
IRL (dB)	13	10	10	25	13
OCP1dB (d Bm)	21	20	22.5	16	24

dependent but fixed, the current through R_3 is fixed. Resistor R_6 is needed to ensure that a small amount of current will flow through Q_1 and thus maintain bias stability. The LNA will in this case be realized as a multi-module IC (MMIC) with a few additional miniature surface-mount discrete components in an encapsulated chip-on-board (COB) package. The resulting module needs only a few external standard decoupling capacitors.

A typical example for an LNA with integrated active bias control is the ADL5521 series from Analog Devices [2]. At 1950 MHz, it offers a 1 dB NF, +35 dBm OIP3, and a +21 OCP1dB for 300 mW. The IRL can be tuned to the right value with an additional external LC matching network. Also companies like RFMD and Avago are providing LNA solutions for cellular base stations [3,4]. An overview is provided in Table 3.1.

3.2.2 Cascode LNA

The singl-device common-source (CS) topology can be improved using an additional common-gate (CG) structure on top of the CS structure. Such a structure is commonly known as a cascode topology. An example is given in Figure 3.2 [5]. An external inductor at the input node provides noise and input return loss match. Furthermore, an external "choke" inductor at the output provides the DC coupling of the supply voltage to the cascode structure, as well as providing the necessary impedance conditions to have a good output return loss. Capacitor C_1 ensures that the CG-GaAs device is properly grounded for RF, whereas R_1 and R_2 set the correct bias voltage at the gate at DC. Inductor L_e provides additional design freedom to enable simultaneous noise and power match at the input.

When a cascode LNA is implemented in a 0.5 μm GaAs pHEMT technology, a noise figure below 0.5 dB can be obtained in the 600–1100 MHz cellular frequency band with a gain around 20 dB at 1100 MHz [5]. At 900 MHz (GSM band) the output linearity is better than +30 dBm for 300 mW with an output compression point of +22 dBm.

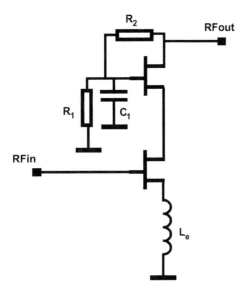

FIGURE 3.2

GaAs cascode LNA (biasing circuitry for CS device not shown).

3.2.3 Cascaded LNA

A cascaded design allows for the adjustment of the bias operating point, and so the noise and linearity performance, at each stage. The first stage can now be optimized for noise while the second stage will ensure that the linearity is maintained. For instance, suppose that for the first stage the NF, power gain, and OIP3 are 0.5 dB, 13 dB, and +30 dBm, respectively. Similarly for the second stage assume 1.5 dB, 10 dB, and +36 dBm for NF, power gain, and OIP3, respectively. Then the overall LNA characteristic yields a 0.6 dB NF and an OIP3 of +34 dBm (assuming a 50 Ω environment).

Source pull data on a pHEMT device indicates that the maximum OIP3 can be achieved when the source impedance is smaller than 50 Ω. An inter-stage matching network is therefore needed to achieve the correct input source termination for the second stage. In a two-stage design, as the gain of the first stage increases, to reduce the noise impact of the second stage, the overall compression point starts to degrade due to earlier saturation of the second stage. Typically the input power at the second stage is around −7 to −3 dBm.

The first stage is optimized for the lowest NF, while still achieving the linearity requirement and compression point. Normally inductive degeneration is applied for simultaneously optimizing NF and input match, as we have seen in Figure 3.2. The degeneration is done by the use of a bond wire, giving a much higher-quality factor than an integrated inductor. A low-quality factor of the degeneration inductor would degrade the noise figure by a few hundreds of milli-dB. To achieve low inductance values, many bond wires can be placed in parallel (often placed as down bonds).

A typical example is the product SKY65040-360LF from Skyworks, an LNA designed to operate between 1.5 and 2.4 GHz [6]. At 1.95 GHz it achieves an NF of 0.65 dB for the maximum (adjustable) gain of 25 dB. The OIP3 and OCP1dB are 35 dBm and 16 dBm, respectively, for 65 mA at 5 V supply voltage. The IRL is better than 20 dB. Similar performance numbers are obtained by the MGA-14516 from Avago [7].

An overview of several GaAs-based LNAs for a 1.95 GHz WCDMA base station application is provided in Table 3.1.

3.2.4 **MMIC LNA module**

An extremely low-noise, high-linearity, high-power amplifier can be realized using a combination of a 0.5 μm GaAs e-pHEMT process, a few miniature surface-mount discrete components and an encapsulated chip-on-board package [8]. External 3 dB hybrid couplers enable the balanced operation of the amplifier, see Figure 3.3.

Each gain block can be based upon a cascaded topology of two stages, where each stage is configured as a common-source topology (similar to Figure 3.2). The gain block will be used in a balanced configuration and therefore the input matching is of lesser importance. Consequently, the first stage in the gain block can optimally be configured for best noise figure performance. The second stage in the gain block is, as discussed in the previous section, designed for linearity. Single-ended performance of the gain block including matching at input and output reveals an NF of 0.6 dB at 2 GHz and an OIP3 of +46 dBm.

Connected as a balanced amplifier, the overall performance is also determined by the used hybrids. Typical insertion loss for a 3 dB hybrid ranges from 0.1 to 0.3 dB in the 1.8–2.2 GHz frequency band. At the input of the balanced amplifier, this would cause a dB-for-dB degradation in the NF. At the output, due to the 3 dB combining

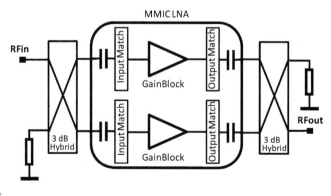

FIGURE 3.3

Block diagram of a balanced MMIC LNA including two external 3 dB hybrids.

effect of the coupler, the OCP1dB and OIP3 should theoretically be increased by 3 dB. The implemented version of [8] shows an OCP1dB of +31 dBm (i.e. beyond 1 W) and an OIP3 is +46 dBm, meanwhile the NF is only 0.9 dB. The power consumption is 4 W indicating that the power-added efficiency of the MMIC LNA is still around 25%.

3.3 Silicon-based LNAs

Recently, silicon-based LNA solutions for wireless infrastructure applications became available. For a long time, III–V compound technology devices were superior to silicon-based devices in terms of noise and linearity performance. In 2000 a 0.25 μm Si BiCMOS LNA achieved a 1.35 dB NF and +26 dBm OIP3 at a moderate gain level of 13 dB [9]. In 2002 a 0.5 μm BiCMOS LNA achieved an NF of 1.4 dB, an OIP3 of +25 dBm, and 16 dB of gain [10]. The performances of these silicon-based LNAs were clearly not in line with the GaAs-based counterparts. But the designs made the advantage of a silicon-based solution clear, e.g. denser integration, higher yield, and the possibility to integrate ESD protection.

The minimum noise factor, F_{min}, of a common-emitter NPN can be approximated by

$$F_{min} = 1 + \frac{1}{\beta} + \sqrt{2R_B g_m \left(\frac{1}{\beta} + \frac{f^2}{f_T^2} \right)},$$

where β is the current gain factor, R_B the base resistance, g_m the transconductance, and f_T the cut-off frequency [11]. Therefore, a technology node with high β and f_T is preferred to realize low-noise factors. Modern SiGe:C BiCMOS processes have cut-off frequencies beyond 200 GHz and have a current gain close to 2000, which almost eliminates the base current shot noise and makes the device rather similar to a GaAs FET in terms of noise performance. The resulting noise performance with NF_{min} lower than 0.5 dB for cellular frequencies is close to that of state-of-the-art GaAs FETs. Table 3.2 gives an overview of two 0.25 μm SiGe:C BiCMOS technologies from NXP Semiconductors. QuBIC4X originates from around 2004, while QuBIC4Xi was developed around 2008 [12–14]. BiCMOS technologies with smaller features size, i.e. 130 nm or 90 nm, do exist and report f_Ts beyond 300 GHz.

A typical f_T versus collector-emitter current plot for the low-voltage NPN in QuBIC4X is shown in Figure 3.4. In the same plot the maximum frequency for which the NF_{min} is still 0.7 dB (fNFmin0.7dB) is also depicted. The plot indicates that within the cellular frequency bands noise figures below 0.7 dB can still be realized. For the NPN in this technology an optimal current density to achieve fNFmin0.7dB or better is 0.25 mA/μm². At this current density the f_T is still beyond 30 GHz or 10 times the operational frequency. This factor 10 is considered to be a safe margin for product manufacturing.

Table 3.2 Comparison of Two Different 0.25 μm SiGe:C BiCMOS Processes from NXP. LV Reflects a High-Performance Low-Voltage Device, HV Refers to the High-Voltage Device

Parameter	QuBIC4X		QuBIC4Xi		Unit
	LV NPN	**HV NPN**	**LV NPN**	**HV NPN**	
β	400	320	2000	1700	
f_T	130	60	216	80	GHz
f_{max}	180	120	177	162	GHz
$NF_{min}@$ 2GHz	0.60	–	0.35	–	dB
BV_{ceo}	1.8	3.3	1.4	2.5	V
BV_{ceo}	6	13	5.2	12	V

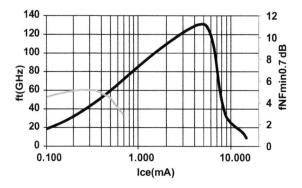

FIGURE 3.4

Cut-off frequency (black) and fNFmin0.7dB (gray) as a function of the collector-emitter current. Emitter area of the NPN device is $1.2\,\mu m^2$.

3.3.1 Single-stage BiCMOS LNA

A bipolar single-stage cascode topology, also referred to as a common-emitter (CE), common-base (CB) topology, is shown in Figure 3.5. Transistor Q_1 is configured as the common-emitter device, and Q_2 as the common-base device. The LNA is inductively loaded by L_1 and to have simultaneous power and noise matching towards the source impedance inductive degeneration by means of L_e has been applied. In fact this is a similar technique used in the GaAs LNA of Figure 3.2. Additionally feedback capacitor C_{fb} is added to improve the power matching at the input. Both bipolar transistors are resistively biased (R_1 and R_2).

A cascode topology can easily provide more than 15 dB gain, but has the drawback that the output impedance is rather high, making matching to a 50 Ω load difficult.

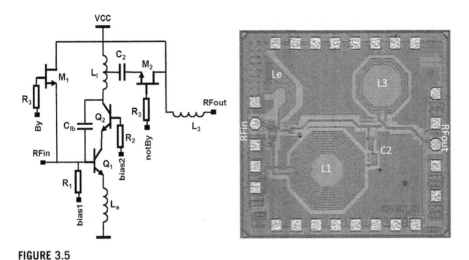

FIGURE 3.5

Circuit diagram for a cascode-topology (left, biasing circuitry not shown) and the die photograph (right).

To overcome this problem, an impedance transformation can be implemented by means of a center tap on load inductor L_1.

In this specific example, a bypass mode has been implemented. In this mode ("By" becomes "high"), the LNA is turned off and the input signal at RFin is routed via M_1 towards output RFout. Inductor L_3 is used in bypass mode to tune out the base-emitter capacitance of Q_1, seen through M_1 and which is then unbiased. Inductor L_3 will also contribute to the output matching in active mode, together with C_2.

As in bypass mode the overall linearity of the LNA can be disturbed by M_1, the influence of the nonlinear device capacitances, i.e. drain-to-bulk and source-to-bulk capacitances, need to be reduced. This can be realized by surrounding the device by deep-trench isolation (DTI) and by connecting the bulk silicon outside this DTI region to ground. A very high resistive path from the back gate of M_1 to ground can be realized in this way. In fact, thanks to the DTI, the local substrate emulates the substrate of an SOI technology.

This topology has been used to realize an LNA for a 1.95 GHz WCDMA base station application [15].

The LNA has been realized in the 0.25 μm SiGe:C BiCMOS technology of NXP (QuBIC4Xi). The die photo is shown in Figure 3.5 (right). For ease of understanding the most important passive components are indicated on the die photo. The measured NF at room temperature for the cascode LNA is given in Figure 3.6, indicating an NF of 0.6 dB at 2 GHz. The power consumption is 290 mW from a 5 V supply in active mode. Other measured performances are provided in Table 3.3, indicating that this BiCMOS design can compete with single-device GaAs-based LNA as discussed in the previous section.

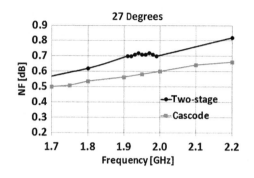

FIGURE 3.6

Measured NF for two different versions of the first gain block as a function of frequency at 27 °C: two-stage topology (black circles) versus cascode topology (gray squares).

Table 3.3 Measured Performance for Two Different LNAs at 27 °C in the 1.95 GHz WCDMA Band

Performance	Single-Stage LNA		Cascaded LNA	
	Active Mode	**Bypass Mode**	**Active Mode**	**Bypass Mode**
Pdiss (mW)	290	0	350	0
Gain (dB)	17	−1.1	19	−1.1
NF (dB)	0.6	1.1	0.7	1.1
IIP3 (dBm)	+8	+41	+14	+41
OIP3 (dBm)	+27	+40	+33	+40
OCP1dB (dBm)	+15	>+15	+15	>+15

3.3.2 Cascaded BiCMOS LNA

The single-stage, cascode LNA lacks very high OIP3. In order to improve the linearity a similar approach as in the GaAs LNA designs can be followed, i.e. the use of a cascaded or two-stage design.

A possible solution of a cascaded LNA is shown in Figure 3.7.

The first stage of the cascaded design is built around the inductively degenerated device Q_1, which is inductively loaded by L_1. The current density in Q_1 is chosen to get the best compromise between linearity, input matching, and noise. Shunt capacitor C_{c1} together with feedback capacitor C_{f1} and series inductor L_{e1} ensure proper input matching. To obtain a good impedance match between the two stages, the output signal of the first stage is taken from the center tap of coil L_1 rather than directly from the collector of Q_1. The second stage is realized by Q_2, where again inductive degeneration is used to improve the impedance matching and linearity.

Special care is taken in the design of the bypass loop. Even in bypass mode the NF may not be degraded too much, putting stringent demands on the insertion loss

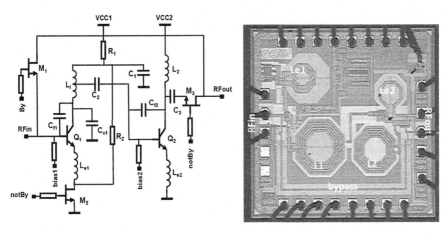

FIGURE 3.7

Circuit diagram for two-stage LNA (left, biasing circuitry not shown) and the die photograph (right).

of the switch M_1 in the loop. In addition, the linearity has to be high in this mode. The bypass path is realized by device M_1. In bypass mode, the emitter of Q_1 will be pulled up via R_2, since M_2 is switched off and the output node of the second stage will be disconnected from the RFout node by means of M_3. This will ensure a proper disconnection of the actual first stage from the bypass loop.

A dedicated LNA for WCDMA base station application in the 1.95 GHz band has been realized [16]. The die photo is shown in Figure 3.7 (right). Again for ease of understanding the most important passive components are indicated on the die photo. The cascaded LNA consumes 350 mW from a 5 V supply. In the 1.92–1.98 GHz frequency band, the measured noise figure in gain mode is 0.7 dB at room temperature (see Figure 3.6), which is 100 mdB higher than the NF of the cascode LNA. Other performance results at 27 °C are provided in Table 3.3. The output linearity has been improved to +33 dBm, a number close to the GaAs cascaded LNA solutions. Note that the 1.1 dB insertion loss in the bypass mode reflects a good NF in this mode.

3.3.3 Broadband BiCMOS LNA

Rather than tailoring an LNA design to a typical cellular communication band as has been done for the designs just discussed, one can also try to realize an LNA operating over a broad frequency range. The starting point of the design in this case is the DC voltage gain of a single transistor gain stage with inductive loading. Like in the previous designs, a DC gain of 17 dB seems reasonable. From Figure 3.4 one can see that optimal NF performance can be expected for a current density of $0.25\,\text{mA}/\mu\text{m}^2$ if using a QUBIC4X technology. Choosing an emitter area of $4\,\mu\text{m}^2$ will give an optimal noise current of 1 mA but would require a large impedance Z_{load} of approximately $1500\,\Omega$ for 17 dB gain. Increasing the current to 1.3 mA for the same device size will hardly impact the noise, but reduces the load impedance to $1200\,\Omega$ or an equivalent inductance of 100 nH (at 2 GHz). From simulations this initial design reveals an NF_{min} of about 0.5 dB with $Z_{opt} = 0.7 + j0.03$. The latter means that $Re(Z_{opt}) = 300 \, \Omega$. Furthermore, G_{max} is in the order of 15 dB and the gain stage has an OIP3 ≈ -12 dBm. On the other hand, $Z_{in} = 18 - j4$, thus there is no simultaneous input impedance match and noise match. The linearity of the initial design can be improved by setting multiple of these stages in parallel. This will obviously increase the total current, but at the same time Γ_{opt}, Z_{in}, and Z_{load} will be decreased by a factor equal to the number of stages in parallel. A number of 12 parallel stages seem a reasonable balance between the total current of 16 mA and OIP3 of +27 dBm.

A cascaded design is needed to improve the overall linearity. As discussed earlier, for a sub-1 dB noise figure the noise contribution of the second stage remains important. If an overall NF of 0.6 dB is the target, the Friis formula would require an NF of 0.7 dB for the second stage, assuming a realized available gain of 15 dB in the first stage. A design strategy similar to the one used for the first stage can be applied resulting in a design with 30 stages in parallel, each stage using a high-voltage device with an emitter length of 20 μm. The resulting OIP3 of this second stage is +40 dBm (including emitter degeneration) while the NF is around 0.7 dB. The main challenge is now in the inter-stage impedance matching when combining the two stages. The output impedance of the first stage is close to 50 Ω, whereas the input impedance of the second stage is in the order of 3 Ω. Similar to the design in Figure 3.7, tapping the inductive load of the first stage helps to lower the output impedance of the first stage without too much performance degradation. Adding a shunt capacitor, C_1, at the collector node will improve the input matching. The tapped inductor can be realized as a 9 nH symmetrical inductor with the center tap as output node.

A die photo of the fully integrated cascaded LNA can be seen in Figure 3.8, where the most important passives have been indicated. Excellent NF performance has been obtained from a packaged sample [17]. The NF is 0.75 B in the 900 MHz GSM band and 0.9 dB in the 1800 MHz band (Figure 3.9). The measured NF and NF_{min} indicate that an almost optimal noise match at the input has been obtained. Meanwhile a very good input impedance match is also obtained as can be seen from the measured input

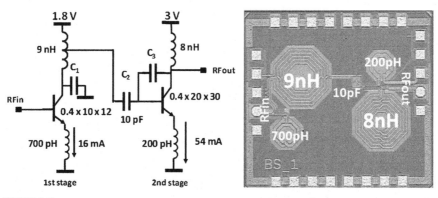

FIGURE 3.8

Broadband LNA design (left, without biasing circuitry) and die photograph (right).

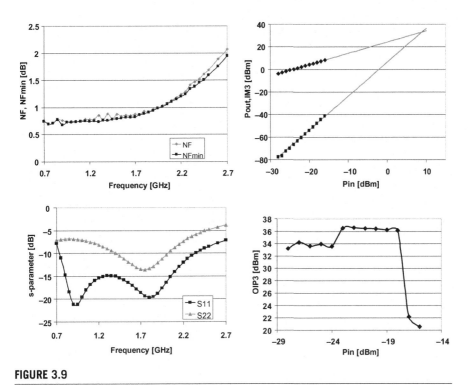

FIGURE 3.9

Measured data on a packaged sample of the broadband LNA: NF and NF_{min} (left top), *s*-parameters (left bottom), 2-tone measurement (right top), and OIP3 (right bottom).

return loss, i.e. S_{11}. The IRL is better than 18 dB at the GSM frequencies. The input impedance match is better than -10 dB between 750 MHz and 2.4 GHz indicating wideband impedance matching. The ORL is better than 7 dB in the 900 MHz and above 10 dB in the 1800 MHz band. The measured transducer gain, i.e. S_{21}, shows the expected 12 dB gain drop per decade, i.e. 6 dB per gain stage.

The linearity measurements have been performed with two tones centered in the 900 MHz band and spaced 80 MHz apart. The measurements reveal an OIP3 of $+36$ dBm and a corresponding IIP3 of $+10$ dBm derived from the extrapolation of the IM3 and gain curve. One can also calculate the OIP3 for each input power separately (Figure 3.9). The measured data reveals an average OIP3 of $+35$ dBm and indicates an ICP1dB of -17 dBm or equivalent to an OCP1dB of $+19$ dBm. A linearity test with tones at 1800 MHz shows similar performance values.

3.3.4 CMOS LNAs with sub-1 dB NF

Like in bipolar designs, CMOS-based LNA designs also suffer from the intrinsic low breakdown voltages and supply voltages to realize high-linearity and high compression point values. However, the NF_{min} of an MOS device can be very low as the gate length is scaled downwards. A 65 nm MOS device has an NF_{min} of 0.2 dB at 2 GHz. The problem is the associated optimal noise impedance and input impedance. The input impedance is mainly capacitive, but needs to be matched to a 50 Ω source impedance. A series inductance at the gate can be used to resonate out the capacitance, but the series resistance of the inductor (implemented as bond wire or integrated on silicon) will spoil a sub-1 dB noise figure.

The first sub-1dB NF LNA in CMOS was presented in the year 2007 and was meant to operate in the 700–1400 MHz frequency band [18]. The topology choice was a single-stage, cascode configuration and included a noise matching circuitry between the common-source and common-gate transistors.

Realized in a 0.18 μm CMOS node, the LNA achieved an average of 0.5 dB NF at 1400 MHz for a 50 mW power consumption. Output IP3 is 15 dBm and output compression point is 6 dBm, indicating that it is very difficult to achieve high linearity and low noise figures at the same time in a CMOS process. The NF is on par with bipolar solutions but the low breakdown voltage of CMOS hampers good large signal performance. The achieved OCP1 dB is almost 10 dB lower than what is commonly achieved in bipolar circuits. The design has been optimized for noise performance in a 90nm CMOS process [19]. For a non-50 Ω input match an NF below 0.2 dB was achieved for a 15 dB gain. The trade-off was in the linearity performance: the OIP3 dropped to 12 dBm, and OCP1dB to 2 dBm.

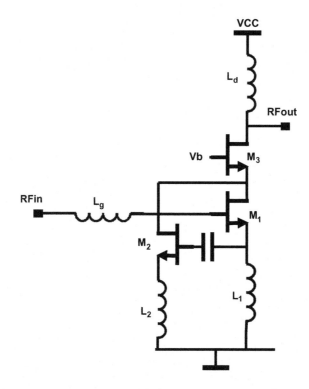

FIGURE 3.10

CMOS LNA with improved linearity (bias circuitry not shown).

The linearity performance and large signal performance can be improved using an auxiliary signal path [20]. The transistors M_1 and M_2 (see Figure 3.10) form a basic cascode configuration with inductive loading by L_d and inductive degeneration by L_1. Transistor M_2 is also inductively degenerated by L_2 and is used to tune the magnitude and phase of the IM3 component. The nonlinearity information in the drain current of M_1 is tapped off as voltage at the source of M_1 and is fed into M_2. This device as well as the inductor L_2 is tuned such that they produce IM3 components which can cancel out the IM3 component produced by M_1. This information is fed into the signal path as current.

Realized in a $0.35\,\mu m$ CMOS process, the LNA achieves an IIP3 of $21\,dBm$ for $11\,dB$ gain. The NF, however, is almost $3\,dB$ at $900\,MHz$. The OCP1dB is $1\,dBm$ indicating that the technique mainly helps to improve the small-signal linearity but not the large-signal performance.

CONCLUSIONS

In this chapter several low-noise amplifier topologies in various technologies have been discussed. The intended application is the cellular base station, where the combination of a sub-1 dB noise figure and high linearity sets interesting design challenges. In Table 3.4 several discussed LNAs are put together for comparison.

The table shows that amplifiers realized in state-of-the-art BiCMOS technologies can achieve similar noise and linearity performances as their GaAs counterparts. However, the GaAs solutions still outperform the silicon-based solutions in the achieved output compression point, thanks to the higher breakdown voltage in GaAs processes.

Currently, CMOS amplifiers cannot offer the combination of sub-1 dB NF (for good IRL) and high linearity and high compression point. Where the breakdown voltage of modern high-performance NPN devices is around 1.8 V, deep sub-micron MOS devices have to cope with breakdown voltages around 1 V. This makes an ICP1dB beyond 0 dBm challenging.

The MMIC in [8] shows excellent performances but the 4 W power dissipation is too much for a tower-mounted base station LNA.

Table 3.4 Comparison of Several LNA Designs for 1.9–2.0 GHz Operation

Performance @ 1950 MHz	[3]	[6]	[8]	[14]	[15]	[16]
Configuration	Single stage	Cascaded LNA	MMIC LNA	Single stage	Cascaded LNA	MMIC LNA
Pdiss (mW)	450	325	4000	290	350	190
Gain (dB)	15	25	21	17	19	12
NF (dB)	0.8	0.7	0.9	0.6	0.7	0.9
OIP3 (dBm)	35	35	46	27	33	36
IRL (dB)	10	25	19	20	20	18
OCP1dB (dBm)	22.5	16	31	15	15	19
Technology	GaAs	GaAs	GaAs	BiCMOS	BiCMOS	BiCMOS

References

[1] 3GPP TS 25.104 v10.2.0 (2011-06): 3rd Generation Partnership Project; Technical Specification Group Radio Access Network; Base Station (BS) radio transmission and reception (FDD) (Release 10).

[2] Analog Devices, ADL5521, 400 MHz to 4000 MHz Low Noise Amplifier, Data Sheet.

[3] RFMD RF3863: Wide Bandwidth, High Linearity Low Noise Amplifier, Data Sheet.

[4] Avago Technologies, ATF-54143: Low Noise Enhancement Mode Pseudomorphic HEMT in a Surface Mount Plastic Package, Data Sheet.

[5] J. Staudinger et al., Wide bandwidth GSM/WCDMA/LTE base station LNA with ultra-low sub 0.5 dB noise figure, in *Radio and Wireless Symposium (RWS)*, pp. 223–226, 2012.

[6] Skyworks, SKY65040-360LF: Low Noise Amplifier 1.5–2.4 GHz, Data Sheet.

[7] Avago Technologies, MGA-14516: High Gain, High Linearity Active Bias Low Noise Amplifier, Data Sheet.

[8] T. Chong, A low-noise, high-linearity balanced amplifier in enhancement-mode GaAs pHEMT technology for wireless base-stations, in *Proceedings of the Gallium Arsenide and Other Semiconductor Application Symposium, EGAAS*, pp. 461–464, 2005.

[9] O. Boric-Lucbeke et al., Si-MMIC BiCMOS low-noise high-linearity amplifiers for base-station applications, in *Proceedings of the Asia-Pacific Microwave Conference*, pp. 181–184, 2000.

[10] V. Aparin et al., Highly linear SiGe BiCMOS LNA and mixer for cellular CDMA/AMPS applications, in *Proceedings of the RFIC*, pp. 129–132, 2002.

[11] G. Niu et al., Noise-gain tradeoff in RF SiGe HBTs, in *Digest of Papers of Topical Meeting on SMIC*, pp. 187–191, 2001.

[12] W. Van Noort et al., BiCMOS technology improvements for microwave application, in *Proceedings of the BCTM*, pp. 93–96, 2008.

[13] P. H. C. Magnee et al., SiGe:C profile optimization for low noise performance, in *Proceedings of the BCTM*, pp. 167–170, 2011.

[14] P. Deixler et al., QUBiC4X: An fT/fmax = 130/140 GHz SiGe:C-BiCMOS manufacturing technology with elite passives for emerging microwave applications, in *Proceedings of the BCTM*, pp. 233–236, 2004.

[15] J. Bergervoet et al., A 1.95 GHz sub-1 dB NF, +40 dBm OIP3 WCDMA LNA, *IEEE Journal of Solid-State Circuits*, vol. 47, no. 7, pp. 1672–1680, 2012.

[16] J. Bergervoet et al., A 1.95 GHz sub 1-dB NF, +40 dBm OIP3 WCDMA LNA with variable Attenuation in SiGe:C BiCMOS, in *Proceedings of the ESSCIRC 2011, Helsinki*, pp. 227–230, 2011.

[17] D. Leenaerts et al., 900MHz/1800MHz GSM basestation LNA with sub-1dB noise figure and +36dBm OIP3, in *Proceedings of the RFIC 2010, Anaheim*, pp. 513–516.

[18] L. Belostotski and J. W. Haslett, Noise figure optimization of wide-band inductively-degenerated CMOS LNAs, in *Symposium Proceedings of the MWSCAS-2007*, pp. 1002–1105.

[19] L. Belostotski and J. W. Haslett, Sub-0.2dB noise figure wideband room-temperature CMOS LNA with non-50 ohm signal source impedance, *IEEE Journal of Solid-State Circuits*, vol. 42, pp. 2492–2502, 2007.

[20] S. Ganesan et al., A highly linear low-noise amplifier, *IEEE Transactions on Microwave Theory and Techniques*, vol. 54, no. 12, pp. 4079–4085, 2006.

High-Efficiency Power Amplifiers for Wireless Infrastructure

4

Mustafa Acar, Mark P. van der Heijden, and Jawad H. Qureshi

NXP Semiconductors, Eindhoven, The Netherlands

INTRODUCTION

Wireless communication is an indispensable part of our daily life in today's world. Use of smartphones, tablet PCs, and other wireless devices/services creates a huge amount of data traffic. With the increasing number of subscribers, expected to be more than 6 billion by 2014 [1], handling high data traffic with limited frequency resources will continue to be one of the prime challenges for wireless communication industries. Transmission of huge amounts of data implies high energy consumption. The global information communication technology (ICT) industry accounts for about 2% of the total human CO_2 footprint, which is comparable to the air traffic CO_2 emissions all around the world [2]. Of the 2% total ICT emission, 25–30% is produced by worldwide telecom infrastructures and devices [3]. Therefore, energy consumption due to telecom infrastructures is significant. Besides, most of the energy in mobile networks is consumed by the base stations [4,5]. Typically more than half of the energy consumption in a base station is due to the RF power amplifier, as can be seen in Figure 4.1. Therefore, improving the efficiency of the power amplifier will decrease the energy consumption and hence reduce the cost of telecom infrastructures.

Increasing the efficiency of the power amplifiers in base stations is very challenging for many reasons. A typical property of the signals in third-generation (3G) and beyond communication systems is high peak-to-average power ratio (e.g. in wideband code division multiple access (WCDMA) typically 10 dB), requiring high linearity of the transmitting amplifier. Consequently, PAs are typically dimensioned for the peak power condition, but are operated most of the time at significantly lower power levels (i.e. power back-off). As a result, even when using a high-efficiency amplifier class, e.g. (inv.) class B [6], class E [7] or (inv.) class F [8], the peak efficiency might be high, but the average amplifier efficiency can be rather low. For this reason, there is a renewed interest in highly efficient PA architectures at back-off power levels that were introduced in the 1950s and 1960s.

Figure 4.2 shows the power amplifier concept evolution in wireless communication systems. The design of power amplifiers in early 2000 was based on linear but

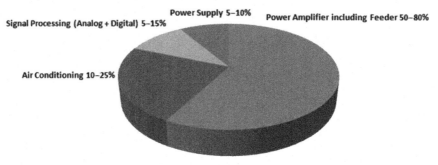

FIGURE 4.1

Breakdown of conventional base station power consumption [1,2].

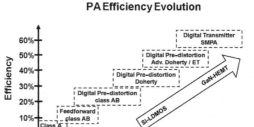

FIGURE 4.2

Power amplifier concept evolution in base stations.

low-efficiency power amplifier topologies such as class A and class AB. Later, use of digital pre-distortion techniques and Doherty power amplifier topology (DPA) has helped to increase efficiency. Further demand for high-efficiency in wireless transmitters requires a shift of the PA topology to digital transmitters, switch-mode power amplifiers (SMPA) such as linear amplification using nonlinear components (LINC) or envelope elimination and restoration (EER).

Most of today's base station PAs use LDMOS technology. Higher RF performance of GaN over LDMOS (e.g. higher supply voltage, higher input/output impedance, etc.) will make it the preferred transistor technology for future base station PAs.

In this chapter, Section 4.1 explains the basics of Doherty PAs by presenting a wideband implementation. Afterwards, an outphasing switch-mode PA will be introduced in Section 4.2. Finally, an essential part of switch-mode power amplifiers, drivers, will be presented in Section 4.3.

4.1 **Wideband Doherty power amplifier**

Currently, the Doherty power amplifier (DPA) [9] represents the most commonly used high efficiency concept in base station applications. Its popularity results from its high-efficiency performance achieved at a low hardware complexity and cost level [10,11]. However, a well-known disadvantage of the DPA is its narrow bandwidth (typically ≤10%) [12,13], which complicates its application in multi-band/multi-standard communication systems. Consequently, custom DPA solutions for each individual need in the market have to be developed, raising costs and yielding logistic problems.

To address this limitation, we investigate how the DPA bandwidth for high-efficiency operation can be expanded. This chapter is organized as follows: after providing a brief introduction to the Doherty power amplifier, we first discuss the DPA bandwidth restrictions in Section 4.1.1. In Section 4.1.1 we evaluate various matching topologies (mainly addressing the wideband compensation of the output capacitance of these devices), connection schemes, and Doherty power combiners that extend the DPA bandwidth. The actual circuit realizations are given in Section 4.1.2, followed by experimental verification in Section 4.1.3.

4.1.1 **Theory of Doherty power amplifiers**

Doherty power amplifier operation is a well-established technique to increase the average efficiency of microwave amplifiers [14]. In this technique the RF amplification is accomplished by using multiple branch amplifiers in parallel. When the power outputs of these branch amplifiers are cleverly combined, active load modulation occurs, which helps to improve the average efficiency of the overall DPA amplifier. Currently, there exist various Doherty power amplifier implementations, e.g. the symmetrical two-way DPA [15], the asymmetrical two-way DPA [11], and the three-way DPA [12]. However, in the following sections only the bandwidth performance of the two-way DPA is discussed in detail for reasons of simplicity. The results obtained from this analysis can be easily extended to cover other more complex DPA architectures. The diagram of a two-way Doherty power amplifier (DPA) in its most elementary form is given in Figure 4.3. Here the main and peak

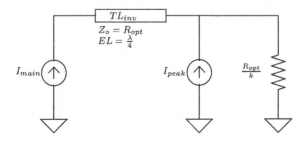

FIGURE 4.3

Schematic illustrating the basic principle of a two-way DPA.

devices are represented by current sources connected by a transmission line impedance inverter, which has an electrical length of $\lambda/4$ at the center frequency of the
design. In the DPA the main device is active at all output power levels, while the
peaking device operates only at the peak power levels [14]. Moreover, the characteristic impedance of the transmission line, TL_{inv}, is chosen in such a way that the
load of the main device (when operating alone) is "x" times its optimum load for
providing maximum power (i.e. $x R_{opt}$). This increased loading results in a higher
RF voltage swing at the drain terminal of the main device, and consequently higher
efficiency of this device [14]. The RF voltage swing and the efficiency of the main
device achieve their maximum value when the output current of the main device
reaches $\left(\frac{1}{x}\right)$ times the current $I_{m,max}$ at full power.

When the RF voltage reaches its maximum (e.g. twice the supply voltages
when operated in class B) the main device gets saturated. As a result, the voltage
swing can no longer be increased. Therefore in order to increase the output power
the load of the main device needs to be decreased. This is accomplished by acti-
vating the peaking device at the point where the main amplifier reaches its saturation voltage. When the peaking amplifier starts to inject current into the load, R_L,
the effective load of the impedance inverter (TL_{inv}) will increase. Consequently, the
main device will experience a decreasing load at its drain terminal due to the impedance inversion. As a result, the main device will come out of saturation and will
increase its output power by injecting more current into the load. In this way, at these
higher power levels, the currents of the main and peaking device(s) are varied such
that the main device always remains at the edge of saturation and hence operates
at maximum efficiency. When reaching full power the peaking device is injecting
($k=x-1$) times more power than the main device, while both devices experience
their optimum load for output power and maximum efficiency. Due to this previously
described operation, the efficiency characteristics of the two-way DPA exhibit two
distinct efficiency peaks. Namely, one at full power where both main and peaking
devices are operating together, and one at the back-off power level where the main
device is operating alone and reaches its voltage saturation.

The difference between a symmetrical and asymmetrical DPA lies in the ratio
of the maximum output power of the peaking amplifier with respect to that of the
main amplifier $\left(k = \frac{P_{peak,max}}{P_{main,max}}\right)$ [10]. This ratio also defines the location of the second
efficiency peak in the efficiency characteristics of the DPA. In a symmetric DPA
the value of k is equal to 1, which results in equally sized main and peaking devices
and a location of 6 dB power back-off level for the second efficiency peak; whereas
for an asymmetric DPA the value of k can be higher, e.g. equal to 2. Note that this
particular choice would result in a peaking device twice as big as the main device
and a 9.6 dB power back-off location for the second efficiency peak. The efficiency
characteristics of a symmetric and asymmetric two-way DPA are shown in Figure 4.4.
In a broader sense, the choice of k will depend on the characteristics of the signal to
be amplified. However, in practical situations the most commonly used values of k
are 1.0 and 2.0 [11].

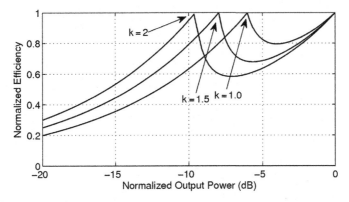

FIGURE 4.4

Theoretical efficiency characteristics of two-way DPA. The various curves show the performance of two-way DPA for different values of *k*.

4.1.1.1 Bandwidth of the DPA with ideal devices

As discussed in the previous section, the operation of the DPA depends on the inversion properties of the impedance inverter transmission line, TL_{inv}, which, besides being vital for the operation of the DPA, is also the only frequency-dependent component in this elementary DPA schematic (Figure 4.3), when implemented using ideal current sources like PA devices. Optimum wideband operation of the DPA would require the impedance inverter transmission line to be a perfect impedance inverter over the whole desired frequency band. Without changing the physical properties of the TL, this is not possible [16]. The electrical length of the TL increases linearly with frequency; hence, the TL forming the inverter will impose bandwidth restrictions on the DPA concept. In light of this, the phase relations versus frequency of the input signals, which control the main and peak device current sources, should track perfectly the phase delay provided by the TL impedance inverter at each frequency [17]. In practical situations, the output capacitance of the PA devices proves to play a very decisive role in defining the maximum bandwidth of the DPA in addition to the impedance inverter, something that we will include later in our analysis.

To evaluate the frequency-dependent efficiency behavior of the DPA, we use the simplified schematic of the two-way DPA with the assumption that the PA devices can be represented by ideal current sources with zero output capacitance. Moreover, the terminal voltages of PA devices (which are vital for efficiency) can be calculated easily if the network enclosed by the dotted rectangle in Figure 4.5 is represented by the impedance matrix (Z-matrix) [16],

$$
\begin{bmatrix}
\dfrac{Z_o\left(R_L\cos(f_n\frac{\pi}{2})+jZ_o\sin(f_n\frac{\pi}{2})\right)}{Z_o\cos(f_n\frac{\pi}{2})+jR_L\sin(f_n\frac{\pi}{2})} & \dfrac{Z_oR_L}{Z_o\cos(f_n\frac{\pi}{2})+jR_L\sin(f_n\frac{\pi}{2})} \\[4mm]
\dfrac{Z_oR_L}{Z_o\cos(f_n\frac{\pi}{2})+jR_L\sin(f_n\frac{\pi}{2})} & \dfrac{Z_oR_L\cos(f_n\frac{\pi}{2})}{Z_o\cos(f_n\frac{\pi}{2})+jR_L\sin(f_n\frac{\pi}{2})}
\end{bmatrix}.
\tag{4.1}
$$

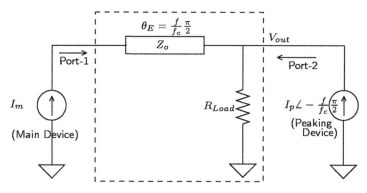

FIGURE 4.5

Basic schematic of the Doherty power amplifiers used in the bandwidth analysis.

Note that in (4.1), f_n is the normalized frequency with respect to the center frequency of the design $\left(f_n = \frac{f}{f_c} \right)$, Z_o is the characteristic impedance of the TL impedance inverter, and R_L is the load impedance. The terminal voltages of the main (V_m) and peaking amplifier (V_p) are given by

$$V_m = Z_{11} I_m + Z_{12} I_p, \tag{4.2}$$

$$V_{\text{out}} = V_p = Z_{21} I_m + Z_{22} I_p. \tag{4.3}$$

I_m and I_p in (4.2) and (4.3) represent the currents of the main and peaking devices. Once the terminal voltages are known (by using (4.1)–(4.3), the output power (P_{out}) and the efficiency (η_{dpa}) of the total DPA, (assuming class-B operation for the active devices) can be calculated as,

$$\eta_{\text{dpa}} = \frac{P_{\text{out}}}{P_{\text{dc}}} \rightarrow \frac{\pi}{4} \frac{Re(V_m I_m^*) + Re(V_p I_p^*)}{|V_{m(\text{max})}||I_m| + |V_{p(\text{max})}||I_p|}. \tag{4.4}$$

Note that class-B operation is assumed in (4.4). The resulting efficiency curves are given in Figure 4.6 for the symmetrical two-way Doherty case.

Note that for ideal devices, which act like perfect current sources, the CW efficiency versus frequency at 6 dB power back-off (see Figure 4.4) indicates that Doherty operation in its most elementary implementation allows an efficiency bandwidth of 28% (e.g. bandwidth where the efficiency is within 10% of its maximum value), while at full power there seems to be no bandwidth restriction. Note that bandwidth remains the same for deeper back-off power levels ($\geq$6 dB for a symmetric two-way DPA) because the main amplifier operates alone, and the loading conditions are identical to the situation at 6 dB power back-off.

The above set of Eqs. (4.1)–(4.4) can also be applied to the asymmetrical Doherty case ($k \geq 1$). The resulting CW efficiency response versus frequency is shown in

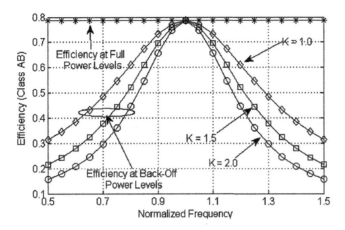

FIGURE 4.6

CW efficiency characteristics of a generic two-way DPA with different values of the asymmetry factor *k*.

Figure 4.6. As expected, the asymmetrical DPA, due to its higher load modulation, shows a narrower bandwidth behavior than the symmetrical DPA. Moreover, in the asymmetrical Doherty the bandwidth reduces with increasing *k*.

Although relatively straightforward to analyze, these results are quite surprising, since conventional DPA implementations show only narrow-band operation, typically resulting in an efficiency bandwidth of only 5–10%. This fact indicates that in conventional designs the TL impedance inverter is not the only bandwidth-limiting component.

In light of this, attention also needs to be given to the bandwidth constraints imposed by other frequency-dependent components in practical DPA implementations (e.g. the output capacitance of the PA devices). Doing so can direct us to new design methods that facilitate higher bandwidths for practical DPA designs.

4.1.1.2 DPA bandwidth for devices with output capacitance

To make our efficiency bandwidth analysis more realistic, the analysis is now extended by including the non-zero output capacitance of the active devices. Because practical PA devices have both output and input capacitance [14], their inherent bandwidth limitations (for a particular reflection coefficient, Γ_m) due to input and output matching are given by Eq. (4.5) for a parallel RC circuit [18,19]

$$\Delta\omega\ln\left(\frac{1}{\Gamma_m}\right) < \frac{\pi}{R_{opt}C_{dev}}, \qquad (4.5)$$

where Γ_m is the reflection coefficient, R_{opt} is the output load, which is usually equal to the optimum load of the PA device for maximum performance, and C_{dev} is the output capacitance of the PA devices.

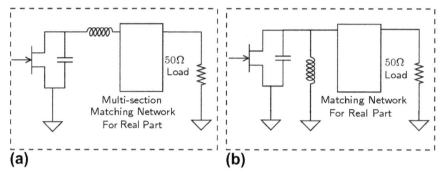

(a)   **(b)**

FIGURE 4.7

(a) Conventional wideband matching; (b) matching topology typically applied in load modulated PA's.

However, this maximum bandwidth can only be achieved if the PA devices are matched using wideband impedance matching techniques, which is traditionally done by matching the output of the PA device as a (complex) impedance [19]. In such an approach, the reactance is first tuned out by a series inductor, after which the remaining real part is matched by a multi-section wideband matching network (Figure 4.7).

Doherty operation, however, requires load modulation [9]; therefore, it is more practical for DPA applications to match the PA devices as an admittance such that the load modulation can be applied directly at the output terminal of the PA devices [14]. This is usually done by first tuning out the output capacitance by a parallel inductor, and then providing the matching for the real part of the output admittance [10] (Figure 4.7). With this topology, the matching of the reactive part forms a parallel resonator at the output of the PA device, which by itself will have a bandpass characteristic and therefore not only limit the bandwidth of the PA devices, but also of the total DPA. The related efficiency degradation of the PA devices, a function of C_{dev} and R_{opt}, can be represented as:

$$\eta_{\text{degrad}} = \cos\left(\tan^{-1}\left(2\pi C_{\text{dev}} R_{\text{opt}}(1 - f_n^2)\right)\right),$$

(4.6)

when deviating from the center frequency of the design.

The resulting efficiency given by (4.6) proves to be related to the output capacitance and optimum loading of the PA device which is characteristic for a given technology.

The effect of this efficiency degradation due to the parallel resonator can also be easily incorporated in (4.1)–(4.4). Moreover, the CW efficiency–frequency characteristics of the DPA with devices having non-zero output capacitance are calculated for NXP's LDMOS Gen7 [11] devices and Cree Inc. GaN devices [20], and are plotted in Figure 4.8. The curves in Figure 4.8 show now the same narrow-band behavior which is usually observed in practical implementations of DPAs.

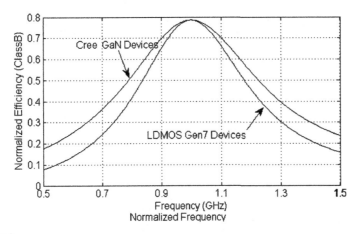

FIGURE 4.8

CW efficiency versus relative bandwidth of a symmetric two-way DPA at 6 dB output power back-off level for two different FET technologies.

Note that the output capacitance of the GaN devices from Cree is almost half the capacitance of the LDMOS devices for a comparable output power level. This reduced output capacitance results in much better bandwidth characteristics of the individual devices (see Eq. (4.6)) thus enabling the design of very wideband amplifiers with this technology. However, in DPA-like configurations this reduced capacitance improves the bandwidth by only a few percent. Therefore, if conventional matching techniques (compensation with parallel inductors) are applied in DPA configuration, the bandwidth cannot be dramatically improved by optimizing the device technology.

It can be concluded from this section that the main cause of the narrow bandwidth of the DPA is the output capacitance of the PA devices and not the commonly blamed output power combiner. Therefore, the first step to improve the bandwidth of the DPA is to compensate the output capacitance of the PA devices in a wideband fashion in order to eliminate the parallel resonating structure at the output of the PA devices [21]. This topic will be discussed in the next section.

4.1.1.3 Output capacitance compensation for wideband operation

One attractive technique to compensate the output capacitance of active devices is to absorb the output capacitance along with the connecting bond-wires in the impedance inverter [21]. Doing so, the efficiency versus frequency behavior improves, since in this situation we are only limited by the bandwidth of the quasi-lumped transmission line, rather than by the bandwidth of the parallel resonator at the output of the active device. Figure 4.9 shows the related DPA schematic, which now also includes the output capacitances and connecting bond-wires of the active devices.

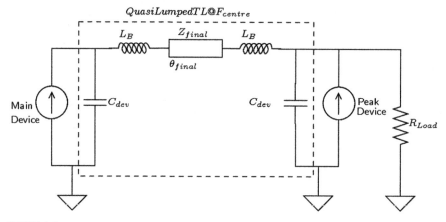

FIGURE 4.9

Simplified schematic of a DPA which absorbs the output capacitance of the main and peak devices, as well their connecting bond-wires into the quasi-lumped transmission line impedance inverter.

Note that the transmission line along with bond-wires and output capacitance of the PA devices, indeed, results in a quasi-lumped TL. Consequently, if the length and characteristic impedance of this artificial TL are adjusted in the proper way, it will act as an impedance inverter at the design frequency.

In order to evaluate the usability and limitation of the proposed concept, an analysis of this quasi-lumped transmission line is performed. For this purpose we ignore the bond-wire inductances in the initial analysis for the sake of simplicity. Doing so, (4.7) and (4.8) give the electrical length and characteristic impedance of the required connecting transmission line needed to absorb the output capacitance of the PA devices in the quasi-lumped inverter

$$\theta_T = \cos^{-1}(\omega C_{\text{dev}} Z_o), \tag{4.7}$$

$$Z_T = \frac{Z_o}{\sin(\theta_T)}. \tag{4.8}$$

Z_o is the characteristic impedance of the original full $\lambda/4$ impedance inverter, while θ_T and Z_T are the calculated electrical length and characteristic impedance of the TL, which forms the inverter along with the output capacitances of the PA devices. Equation (4.8) also imposes a restriction on the maximum capacitance value that can be absorbed in the quasi-lumped transmission line, which for a particular frequency (ω) is given by,

$$C_{\text{dev}} < \frac{1}{\omega Z_o}, \tag{4.9}$$

which in turns sets a limit on the maximum design frequency of wideband DPA using this method,

$$\omega_{max} = \frac{1}{Z_o C_{dev}}. \tag{4.10}$$

As discussed before, Z_o is normally taken to be equal to the optimum load of the active devices (R_{opt}) for delivering maximum power [14]. Therefore, when assuming linear power scaling, the product of Z_o and C_{dev} is fixed for a particular technology and cannot be chosen independently. Consequently, (4.10) limits the maximum design frequency which can be selected for this particular DPA design technique with a given device technology.

Once the length and the characteristic impedance of this TL are known, then (4.11) and (4.12) can be used to refine the length and characteristic impedance of the TL by including the effect of the bond-wire inductances,

$$\theta_F = \tan^{-1}\left(\frac{Z_T}{\omega L_B}\right), \tag{4.11}$$

$$Z_F = Z_T \frac{\sin(\theta_T)}{\sin(\theta_F)}, \tag{4.12}$$

in which L_B is the bond-wire inductance, while θ_F and Z_F are the required electric length and the characteristic impedance of the final transmission line implementation needed to absorb the output capacitances of the active devices and their connecting bond-wires.

4.1.2 Circuit realization

A 20 W wideband DPA was designed to evaluate the forgoing theory. Two bare dice of 10 W NXP Gen 6 LDMOS devices were used for the branch amplifiers operating in class B mode. The design center frequency is set to 1.95 GHz, in order to avoid the frequency limit imposed by (4.10). The required length and the impedance of the final transmission line were calculated using (4.7), (4.8), (4.11), and (4.12). The complete circuit diagram of the DPA along with the biasing lines and harmonic terminations is shown in Figure 4.10.

In this implementation the connection of the peaking amplifier to the load is very important, since the impedance inversion path of the main to peak amplifier should not be partially shared by the DPA connection to the external load. For this reason the output of the peaking device is directly bonded to the load in our design, along with bonding it directly to the quasi-lumped TL (Figure 4.10). The remaining tasks in the circuit design of the DPA are: the wideband output match to the 50 Ω load, and the wideband input match of the PA devices. In our design we chose to have two individual inputs for the DPA in order to allow independent control of the phase relations of the main and peak device to maximize its wideband performance.

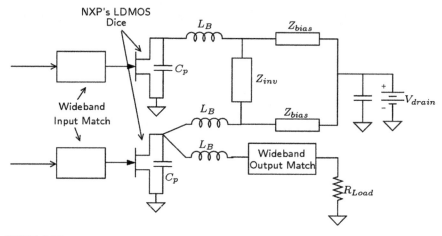

FIGURE 4.10

Simplified block diagram of the 20 W NXP Gen6 wideband DPA demonstrator.

4.1.3 Simulated and measured results

The circuit has been designed using Agilent Advanced System Design (ADS). Figure 4.11 shows the simulation results at maximum output power and at 6 dB power back-off operation. As expected, the DPA has its highest bandwidth at full power while still providing an excellent bandwidth ($\geq$300 MHz) at 6 dB power back-off. The slow efficiency roll-off at full power is due to the frequency behavior of the quasi-lumped transmission, which yields a slightly reactive loading to the PA devices for frequencies higher than the center frequency.

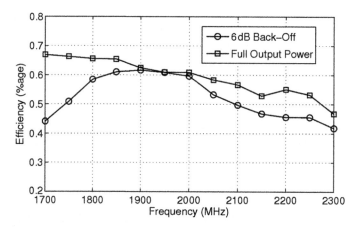

FIGURE 4.11

Simulated efficiency vs. frequency at full output power and at 6 dB power back-off of the 20 W wideband Doherty power amplifier.

Since the goal of this work is to evaluate the wideband DPA performance set by the improved output power combining network, a mixed-signal characterization setup was used to characterize the DPA demonstrator (Figure 4.12). It provides precise control of phase relations of the input signals, such that the output phases of the main and peak device track perfectly with the quasi-lumped impedance inverter [21]. In addition the input signal amplitudes have been optimized for maximum performance. The efficiency measurements are performed from 1.7 GHz to 2.3 GHz, and their results are shown in Figures 4.13 and 4.14. The related gain of the DPA was 13 dB.

The bandwidth limitations of Doherty amplifiers have been evaluated in this section. A wideband output power combing technique for Doherty amplifiers has been

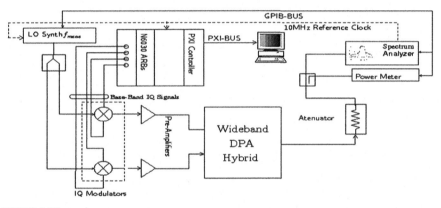

FIGURE 4.12

Block diagram of the measurement setup used to characterize the wideband DPA.

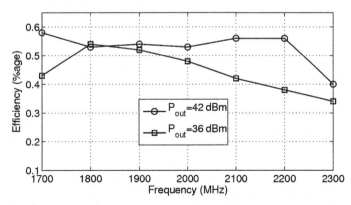

FIGURE 4.13

Measured efficiency at peak output power and 6 dB power back-off versus frequency of the 20 W wideband Doherty power amplifier demonstrator.

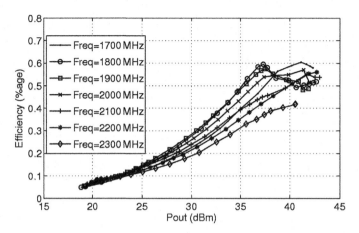

FIGURE 4.14

Measured efficiency vs. output power for the wideband Doherty at different frequencies.

proposed, which has been applied in a two-way Doherty amplifier with independently controlled input signals for the main and peaking devices.

4.2 High-efficiency wideband outphasing power amplifier

A Chireix outphasing PA is a promising candidate to work around the classical linearity–efficiency trade-off and is based on linear amplification using nonlinear components (LINC) [22,23]. In an outphasing transmitter, a complex modulated input signal is split into two signals with constant amplitude and a relative phase difference θ, corresponding to the time varying envelope of the original input signal. As the amplitude is constant the two branch signals can be amplified separately by highly efficient switch-mode PAs (SMPA). After combining both branch signals at the outputs of these SMPAs, an amplified replica of the original input signal results. Unfortunately, due to the non-isolating properties of the combiner, a time-varying reactive load modulation exists at the output of both SMPAs. To mitigate this unwanted load modulation, Chireix proposed compensation elements to be placed at the input ports of the power combiner. This creates an efficiency peak at a specified power back-off level, resulting in an improved average PA efficiency.

Note that a Doherty PA concept requires linear PAs whereas outphasing allows fully saturated or compressed PAs. Therefore, theoretically, an outphasing PA can achieve higher power efficiency than a Doherty PA.

The Chireix outphasing combiner is usually based on quarter-wave transmission lines (QWTL) and can be found in many publications on outphasing PAs [24,25]. The Chireix compensation elements are either lumped or can be incorporated in the combiner [26,27]. Although the classical QWTL Chireix combiner and its asymmetric derivatives look very elegant, there are still some drawbacks. First, the efficiency does not only depend on the outphasing angle, but also on frequency, since both the Chireix compensation elements and the QWTL are frequency dependent. This frequency degeneration is highly undesirable for a wideband outphasing transmitter. Secondly, class-B, -D, and -F implementations have traditionally been used as branch PA, but recently class E has been identified as an even better candidate demonstrating higher efficiency over a wider dynamic range [27].

This section presents a new power combiner topology to circumvent the frequency limitation of the traditional QWTL combiner. The proposed combiner incorporates both the optimum class-E load tuning and the Chireix compensation elements, resulting in a novel class-E outphasing PA with improved efficiency over a wider peak-to-average power range and RF bandwidth [28].

4.2.1 Wideband class-E Chireix combiner

The classical Chireix power combiner is elegant in the sense that both PAs and the load can be single-ended. Ideally, power combining in an outphasing transmitter can be wideband when a floating load is used. The only frequency-dependent elements are the Chireix compensation susceptance (B_C) and the shunt elements that tune the PA for class-E operation (B_E) as shown in Figure 4.15. The PAs are represented by voltage sources to simplify the analysis.

Ideally, a transformer can convert a floating load into a single-ended load [7]. However, a lumped element transformer is difficult to implement for high powers at RF frequencies. Alternatively, coupled lines can be used to combine the outputs like in a Marchand balun (see Figure 4.16). Here we assume symmetrical uniformly coupled-lines to model the balun.

Let us define the general class-E Chireix operating conditions. With the help of Figure 4.15 we can define the prototype admittance conditions Y_1 and Y_2 needed to set up the desired class-E terminations and Chireix compensation elements:

$$Y_{1,2} = j B_{E,LM} \mp j B_C + \frac{G_{E,LM}}{2}(2\sin(\theta)^2 \mp j\sin(\theta)), \quad (4.13)$$

in which

$$B_{E,LM}(q,\omega) = -q^2\omega C_{OUT}|_{q=1.3} \approx -1.69\omega C_{OUT}, \quad (4.14)$$

$$G_{E,LM}(q,k,\omega) = \frac{\omega C_{OUT}}{K_C(q,k)}\bigg|_{q=1.3,k=0} \approx \frac{\omega C_{OUT}}{0.685}, \quad (4.15)$$

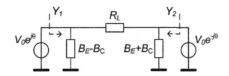

FIGURE 4.15

Ideal class-E Chireix combining.

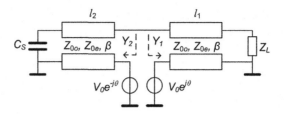

FIGURE 4.16

Coupled-line-based class-E Chireix combiner.

are the optimum class-E susceptance and conductance for load modulation (subscript LM), C_{OUT} is the switch output capacitance, and K_C, q, and k are defined in [29].[1] The required Chireix compensation susceptance is given by

$$B_C = \frac{G_{E,M}}{2} \sin(2\theta_C),\qquad (4.16)$$

in which θ_C is the outphasing angle[2] that determines the efficiency peaking at an appropriate power back-off level.

According to the class-E theory in [29], the parameter q in (4.2) determines how much the resonance frequency of the tank, created by C_{OUT} and the effective shunt-inductance of one coupled line, deviates from the target frequency. In (4.3), k is defined as the slope of the voltage waveform when the switch is turned on and $K_C(q,k)$ is the design equation used for the class-E PA in this work. Among the many possible values, $q = 1.3$ demonstrates an interesting property for an outphasing or load-modulated class-E PA. Unique to this value is that if the load resistance changes from its nominal to a higher value, the class-E PA responds by changing its turn-on voltage slope from zero to a negative value (variable slope) while keeping its turn-on voltage close to zero as it is shown in Figure 4.17. In this way, efficiency is preserved for the varying load conditions that occur in an outphasing PA.

The element values of the class-E Chireix combiner in Figure 4.16 can now be determined by equating the prototype admittance in (4.1) to the input port admittances of the coupled-line combiner. A necessary condition is that at one side the coupled-line section is terminated with capacitor C_S, given as

$$C_S = \frac{Y_{0o} + Y_{0e}}{2\omega} \cot(\beta l_2),\qquad (4.17)$$

which tunes out the leakage inductance of this coupled-line transformer element. The electrical lengths βl_1 and βl_2 of the coupled lines are unequal due to the Chireix compensation elements ($\mp B_C$) and can be calculated with

---

[1]Please note that definition of k is not the same as in the previous section.
[2]Please note that definition of θ is not the same as in the previous section.

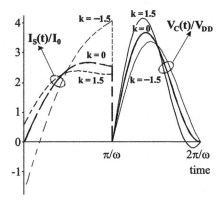

FIGURE 4.17

Variable-slope class-E PA switch voltage waveform.

$$\beta l_{1,2} = \arctan \left(\frac{2}{(-B_{E,LM} \mp B_C)(Z_{0e} + Z_{0o})} \right). \tag{4.18}$$

The last step is to determine the optimum even-mode and odd-mode impedances, Z_{0e} and Z_{0o}. The odd-mode impedance, Z_{0o}, can be calculated based on the impedance transformation from the output (Z_L) to the input port impedances (Z_1 and Z_2):

$$Z_{0o} = \frac{1}{G_{E,LM}} \frac{n - 2 + n^{-1}}{n + 1}, \tag{4.19}$$

in which $n = Z_{0e}/Z_{0o}$, which should be large to maximize the magnetic coupling in the coupled-line transformer. Typically $n \simeq 6-7$.

4.2.2 Class-E RF outphasing power amplifier design

In support of the presented theory, a demonstrator outphasing SMPA has been built with the proposed class-E Chireix coupled-line combiner. Figure 4.18 shows the schematic of the complete PA design. The class-E PA switches are realized in a 28 V GaN HEMT technology. The CMOS drivers are fabricated in standard 65 nm CMOS technology [9]. The drivers are fabricated in a baseline 65 nm CMOS process without any extra mask or process step. The measured f_T of these devices exceed 30 GHz and 50 GHz for PMOS and NMOS, respectively, and their breakdown voltage limit is 12 V [9].

The CMOS drivers are AC-coupled to overcome the negative gate bias requirement of the GaN. Connected to the output of the combiner is a fourth-order Butterworth matching filter that transforms the 50 Ω antenna impedance to the required class-E load and sets the required loaded quality factor.

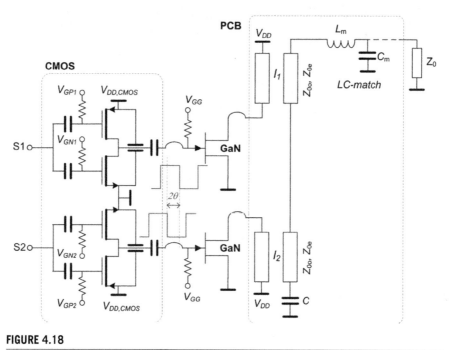

FIGURE 4.18

Simplified schematic of the class-E outphasing PA design.

Figure 4.19 shows the practical implementation of the SMPA module, includ-ing the CMOS drivers, GaN HEMTs, Chireix combiner, and the LC matching filter that connects to the 50 Ω output. The dual CMOS-GaN SMPA line-up is magni-fied for visibility. The Chireix combiner uses symmetrical broadside coupled lines that are implemented between the top metal layers in a dual-layer Rogers laminate with $\varepsilon_r = 3.5$. The top layer is 4 mm thick and nearly levels with the surface of the GaN dies that are attached to the flange directly. The bottom layer is 40 mm thick and terminated with a metal ground plane. The layer thickness aspect ratio basically determines the maximum ratio between the even-mode and odd-mode impedance of the coupled-line implementation.

The GaN SMPA stages need to be driven with pulse wave signals to obtain the highest drain efficiency across all load states. Typically, the input swing is in the order of 5 Vpp to open the GaN switch. Realizing this large signal swing at RF frequencies in a reliable way is a challenge for deep submicron CMOS. In this work, a high-voltage CMOS driver topology is applied using thin-oxide Extended-Drain MOS (EDMOS) transistors. The distance between the CMOS and GaN dies is made small to minimize the bond wire inductances and to maximize the bandwidth of the interface. More detailed information is presented about CMOS drivers in the next section.

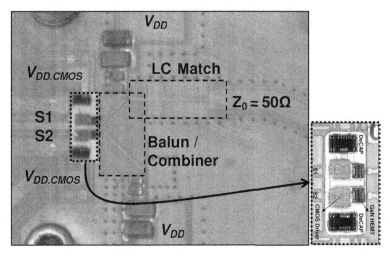

FIGURE 4.19

Photograph of the class-E Chireix outphasing power amplifier.

4.2.3 **Experimental results**

The prototype class-E outphasing SMPA module has been characterized using a dedicated measurement setup capable of acquiring single-tone data and characterizing the PA with complex modulated signals including pre-distortion (See Figure 4.20).

The SMPA module was evaluated by sweeping a single tone from 1.75 GHz to 2.10 GHz. The drain efficiency (DE) is defined as the output power at the 50 Ω connector interface divided by the DC power supplied to the drains of the GaN transistors. The total line-up efficiency also includes the power dissipated by the CMOS

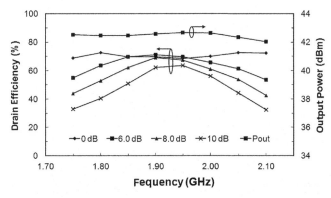

FIGURE 4.20

Measured peak output power and drain efficiency at different power back-off levels as a function of frequency for $V_{DD} = 28$ V, and $V_{DD,CMOS} = 5$ V.

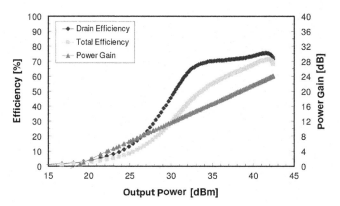

FIGURE 4.21

Measured total line-up efficiency, drain efficiency, and power gain as a function of output power at $f_0 = 1.95\,\text{GHz}$, $V_{DD} = 28\,\text{V}$, and $V_{DD,CMOS} = 5\,\text{V}$.

driver and the RF input power delivered to the inputs of the PA. Figure 4.21 shows the peak power and DE versus frequency at 0 dB, 6 dB, 8 dB, and 10 dB back-off from the peak power level. The peak power varies only 0.7 dB around 42.4 dBm, while maintaining >60% efficiency across >250 MHz at 6 dB back-off, >150 MHz at 8 dB back-off, and >70 MHz at 10 dB back-off. Such performance in terms of efficiency versus bandwidth is remarkably high.

Figure 4.21 shows the drain efficiency, total line-up efficiency, and power gain as a function of output power. At 10 dB back-off, the drain efficiency is 65% and the total line-up efficiency is 44%. At 8 dB back-off, the drain efficiency is 70% and the total line-up efficiency is 53%. Although the proposed SMPA is a two-way system, the drain efficiency at 10 dB back-off is comparable to what has been published for a three-way GaN Doherty PA [10], but now with better wideband capabilities and integration of a CMOS driver stage.

Figure 4.22 shows the SMPA performance when applying a 7.5 dB PAR WCDMA signal after memory-less pre-distortion. The efficiency and ACLR data are listed in Table I (see Figure 4.23), which also contains data for an uncompressed 9.6 dB PAR WCDMA signal.

A design approach toward wideband and high-efficiency RF outphasing PAs is presented in this section. Instead of using the classical quarter-wave transmission line combiner, a wideband coupled-line-based combiner is utilized. Moreover, an advanced class-E mode of operation is proposed for the switch-mode power amplifiers to maintain high efficiency for the time-varying loading conditions during the outphasing operation. Together with the Chireix compensation elements, the class-E terminations are incorporated in an asymmetric coupled-line combiner that acts like a high-frequency transformer. The proposed novel outphasing SMPA concept has been fabricated and measured with realistic complex modulated signals, and demonstrated a state-of-art efficiency versus bandwidth and power back-off ranges.

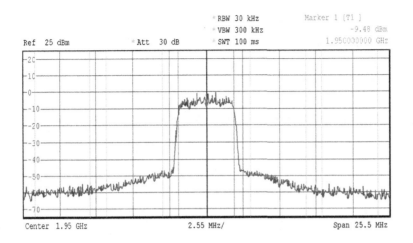

FIGURE 4.22

Measured SMPA output spectrum of a 7.5 dB PAR WCDMA signal after pre-distortion at $f_0 = 1.95$ GHz, $V_{DD} = 28$ V, and $V_{DD,CMOS} = 5$ V.

TABLE I
PERFORMANCE SUMMARY

WCDMA Test signal	P_{max} (W)	$\eta_{DE,av}$ (%)	$\eta_{TOT,av}$ (%)	ACLR1,2 (dBc)
PAR = 7.5 dB	19	65.1	51.6[*]	−47,−54
PAR = 9.6 dB	19	54.5	41.9[*]	−47,−52

[*]including CMOS driver amplifier

FIGURE 4.23

Outphasing PA performance summary.

4.3 High-power drivers for high-efficiency RF power amplifiers

In wireless infrastructure systems, as was mentioned in the previous sections, the high-power PA is often realized in LDMOS or GaN to obtain the required large output powers (e.g. >100 W). Driving these high-power transistors has several challenges.

Typically a GaN or LDMOS PA transistor needs a large voltage swing drive signal, e.g. 5–6 Vpp. Besides the high voltage swing, pulsed drive signals are needed to satisfy high efficiency requirements in wireless infrastructure PAs, e.g. switch-mode

PAs. To preserve the signal pulse shape when driving a high-input capacitance (typically tens of picofarads) of the RF power transistor, a high current is needed. The combination of high voltage and high current at cellular frequencies requires high power capabilities from the PA drivers.

The trend of reconfigurability over a wide bandwidth in wireless transmitters can be enabled by digital techniques for which CMOS is the preferred solution. Moreover, as the driver is the interface between the transmitter and the high-power PA, a CMOS-based driver is advantageous. However, the low breakdown voltage of deep sub-micron CMOS conflicts with the high voltage and high power demands of the PA driver.

In this section, we present two high-voltage RF PA driver designs in baseline 65 nm CMOS. To overcome the breakdown limitations the designs make use of high-voltage extended-drain devices (ED-MOS), which don't require any extra mask or process steps [30,31]. One design, the medium-power (MP) driver, can deliver a pulsed drive signal of 9.6 V peak-to-peak to a 3 pF load capacitance from 0.5 GHz to 4 GHz. It has wide pulse-width controllability and consumes 0.75 W DC power at 2 GHz. The second design, a high-power (HP) driver, can drive a 30 pF load capacitance with 5 W power dissipation at 2.14 GHz.

4.3.1 Driver circuit topology

Figures 4.24a and 4.24b depict the schematic diagram of the MP and the HP driver respectively. Both drivers are mainly targeted for switch-mode PA applications. The ED-MOS devices can be driven directly by low voltage and high-speed standard transistors, simplifying the integration of the output stage with other digital CMOS circuits on a single die. The DC shift required to bias the gate of ED-PMOS is achieved by using a capacitor (C_{IN}) in both drivers.

The MP driver has pulse-width controllability of its output square wave by using the variable-gate-bias technique [32]. Pulse-width control provides a means to perform fine adjustment/tuning functionality to enhance the performance in an advanced switch-mode PA. The bias level at the first inverter stage shifts up/down the RF sinusoidal input signal with respect to the inverter's own switching threshold. A change on this bias voltage will vary the pulse-width at the output of the buffers. Then, this pulse signal will be combined at the ED-MOS inverter of the RF driver.

The HP driver is designed to explore the power upscaling limits of HV ED-MOS inverter for high-power PA applications. By setting the bias voltages at the gates of ED-NMOS and ED-PMOS, the driver can also operate in push–pull Class-B mode, which enables its use for linear wideband PA applications (e.g. wideband Doherty PAs, etc.).

Both drivers have an output AC coupling capacitor (C_{OUT}) which allows a DC shift when driving a power transistor with negative threshold voltage (e.g. GaN).[3]

---

[3]The output voltage, V_{OUT}, is scaled by $C_{OUT}/(C_{OUT} + C_L)$ with respect to V_D due to the presence of C_{OUT}. In this design $C_{OUT} = 10 \cdot C_L$ is chosen, which decreases V_{OUT} by 9.1% with respect to V_D.

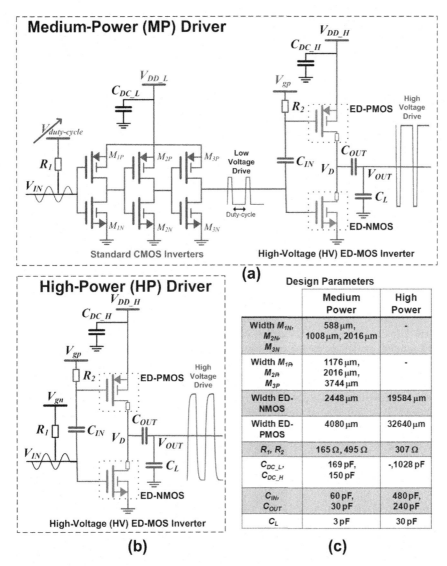

FIGURE 4.24

(a) Schematic of the MP, (b) the HP RF drivers with relevant ideal representation of voltage waveforms, and (c) design parameters.

The following equations are defined by using a generic CMOS inverter in Figure 4.25a in order to evaluate the efficiency performance of the drivers.

$$\text{Efficiency} = \frac{f C_L V_{DD}^2}{V_{DD} I_{DD}}, \tag{4.20}$$

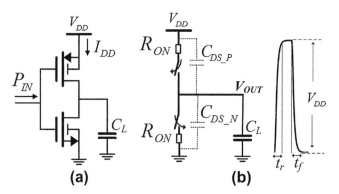

FIGURE 4.25

(a) CMOS inverter and (b) model of CMOS inverter.

$$\text{Total\_Efficiency} = \frac{f C_L V_{\text{DD}}^2}{V_{\text{DD}} I_{DD} + P_{\text{IN}}}. \tag{4.21}$$

In general, switch-mode PAs require drive signals with sharp edges. In the design of high-frequency drivers for switch-mode applications small values of δ (typically <15%, depending on switch-mode power transistor technology) are required to achieve high efficiency, where $\delta = t_r / T = f t_r$. T and f represent the period and frequency. By using the model in Figure 4.25b, we can see the dominant factors influencing δ to achieve in a CMOS inverter design.[4]

It can be seen from (4.22) that δ is proportional to

$$\delta = f t_r \propto f(R_{\text{ON}} C_L + R_{\text{ON}} C_{\text{DS}}), \tag{4.22}$$

where R_{ON} and $C_{\text{DS}} = C_{\text{DS\_N}} + C_{\text{DS\_P}}$ are on-resistance of NMOS and sum of parasitic drain-source capacitance of PMOS and NMOS, respectively. The term $R_{\text{ON}} C_{\text{DS}}$ in (4.22) is a function of technology parameters only [33]. Therefore, for a given technology, C_L and f, a maximum allowed δ is mainly determined by R_{ON}, which depends on the transistor gate width. In general, δ can be decreased by increasing gate width (smaller R_{ON}), which decreases efficiency due to an increase in the power consumption ($V_{\text{DD}} I_{\text{DD}}$) and P_{IN}. If gate width and C_L are constant, δ decreases as f decreases, which is visible in our waveforms as will be explained in Section 4.4. In other words, for a fixed δ a driver can drive higher load capacitance (C_L) toward lower frequencies.

---

[4]In this model, it is assumed that the PMOS gate width is adjusted to have an identical NMOS on-resistance (R_{ON}), resulting in $t_r = t_f$.

4.3.2 **Design and implementation**

Design details of the MP and the HP drivers are given in Figure 4.24c.[5] In the design of the standard three-stage CMOS inverter a minimum possible gate length of $0.06\,\mu\text{m}$ is used. The three-stage inverter is optimized to obtain wide pulse-width control range over a wide bandwidth.

As it is evident from Figure 4.24c, the HP driver is an eight-times upscaled version of the HV ED-MOS inverter part of the MP driver. Many challenges exist in power upscaling of high-frequency drivers. As the power is upscaled eight times, $C_{\text{IN}}, C_{\text{OUT}}, C_{\text{DC}}$, and transistor gate width increase with the same factor. In order to propagate RF pulses via these capacitors without attenuation both self-resonance frequency ($>5\,\text{GHz}$) and quality factor (>10) should be high. Low self-resonance frequency of C_{IN} can cause peaking in the gate-source voltage of ED-PMOS, which may give rise to oxide reliability problems. Besides, the delay in C_{IN} should be minimized to prevent the turning-on of both ED-PMOS and ED-NMOS simultaneously, which will reduce the supply to ground and waste power.

Large-size capacitors and transistors are designed as a parallel connection of unit cells. An important reason for a decrease in self-resonance frequency and quality factor in upscaling the capacitors is parasitic inductance and resistance due to metals connecting unit cells. The connections between the unit cells of capacitors are optimized in order to obtain maximum possible resonance frequency and quality factor using the scalable layout approach that we introduced in [33].

Since the supply and the ground lines of the drivers carry the highest currents ($\approx 1\,\text{A}$) they are implemented using thick wires, metal-6, metal-7, and ALU. The supply decoupling capacitor C_{DC} is distributed all over the chip in order to have a stable V_{DD}. All the capacitors are implemented using custom parallel-plate metal-fringe capacitors.

4.3.3 **Measurement results**

In this section, measurement results of the MP and the HP drivers are presented. The MP driver is measured on-wafer. Rohde-Schwarz SMBV100A Vector Signal Generator is used as the input power source and Agilent E3631A power supplies are used as DC sources. An Agilent DSOX91604A Infiniium 16 GHz Bandwidth Oscilloscope is used to monitor V_{OUT}.

The MP driver is able to operate from 0.5 GHz to 4 GHz, see Figure 4.26. As was explained in Section 4.3.1, δ and efficiency decrease as frequency decreases, which is visible in Figures 4.26 and 4.27a, respectively. The output voltage waveform V_{OUT} has steeper rise and fall at lower frequencies. The Efficiency and the Total_Efficiency are 42% and 33% respectively at 4 GHz. δ and efficiency can be improved by increasing C_L toward lower frequencies.

---

[5]Note that the MP driver has an integrated on-chip load capacitor, $C_L = 3$ pF, whereas the HP driver is measured by driving a GaN power transistor having ≈ 30 pF input capacitance.

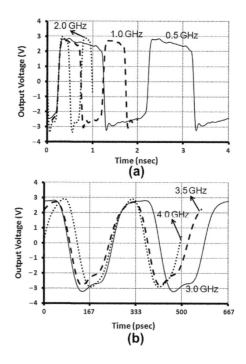

FIGURE 4.26

The MP driver output voltage, V_{OUT}, when $V_{DD\_H} = 6\,V$, $V_{DD\_L} = 1.2\,V$, and $V_{gp} = 5.3\,V$ at (a) 0.5, 1, and 2.0 GHz (b) 3.0, 3.5, and 4.0 GHz. Note that V_{OUT} is biased at 0 V.

DC power consumption of the HV ED-MOS inverter, P_{DC_H}, and the three-stage inverters, P_{DC_L}, increases almost linearly as frequency increases, Figure 4.27b. The measured required input power is very small, maximum 4.2 mW at 4 GHz. There is a close agreement between simulations and measurements in power consumption and efficiency.

The MP driver achieves a pulse-width control (expressed as duty-cycle) of V_{OUT} from 23% to 82% at 1 GHz and from 38% to 73% at 2 GHz, see Figure 4.28. Figure 4.29 shows how duty-cycle depends on $V_{duty-cycle}$ bias voltage at 1, 2, and 3 GHz operation.

Figure 4.30 shows that the MP driver is able to operate with supply voltage (V_{DD_H}) from 3 V to 10 V. In order to see the preliminary reliability performance, the MP driver is operated under 6 V and 10 V supply voltage for 18 h continuously at 1.5 GHz and there was no change in DC bias currents and in V_{OUT} waveform.

The HP driver is tested at 2.14 GHz while driving a 9.6 mm GaN RF power device with ≈30 pF input capacitance, as is seen in Figure 4.31. The power device is connected to a load-pull system and the measured output power is 50 W while the driver consumes 5 W DC power at 6 V supply voltage. The overall power-added efficiency including the driver and the final stage transistor is measured as 63%.

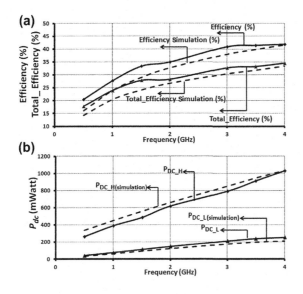

FIGURE 4.27

The MP driver: (a) Efficiency and Total_Efficiency, (b) DC power consumption of the HV ED-MOS Inverter, P_{DC_H} and the three-stage inverters, P_{DC_L}.

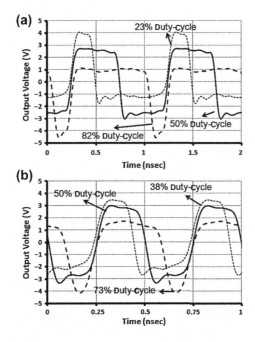

FIGURE 4.28

The MP driver output voltage V_{OUT} for (a) 23%, 50%, and 82% duty-cycle operation at 1 GHz and (b) 38%, 50%, and 73% duty-cycle operation at 2 GHz when $V_{DD\_H}=6$ V, $V_{DD\_L}=1.2$ V, and $V_{gp}=5.3$ V. Note that V_{OUT} is biased at 0 V.

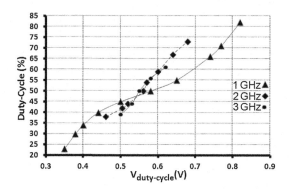

FIGURE 4.29

The MP driver duty-cycle control performance vs. $V_{\text{duty-cycle}}$ at 1, 2, and 3 GHz.

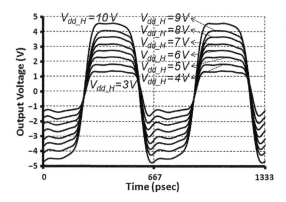

FIGURE 4.30

The MP driver output voltage V_{OUT} at 1.5 GHz when $V_{\text{DD\_H}}$ is increased from 3 V to 10 V by 1 V while $V_{\text{DD\_L}} = 1.2$ V.

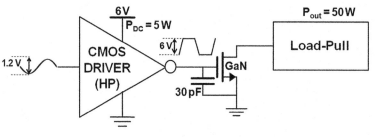

FIGURE 4.31

The HP driver test setup.

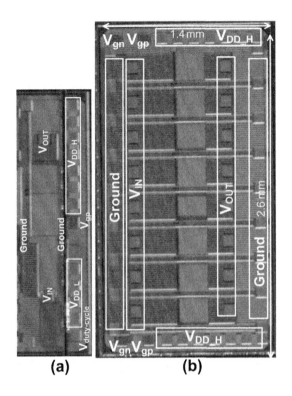

FIGURE 4.32

Chip photo of (a) the MP driver and (b) the HP driver.

Two high-voltage (10 V supply voltage) RF drivers in standard 65 nm CMOS technology are demonstrated (see Figure 4.32). The medium-power driver operates from 0.5 GHz to 4 GHz while driving a 3 pF load capacitance. It consumes 0.75 W DC power at 2 GHz and can achieve wide pulse-width control. The high-power driver consumes 5 W DC power while driving an RF power device (50 W) having ≈30 pF input capacitance at 2.14 GHz.

The CMOS drivers can serve as key building blocks for next-generation reconfigurable multiband multimode transmitters for wireless infrastructure systems, interfacing digital CMOS circuitry with high-power transistors [34].

References

[1] J. Rebello, *Global Wireless Subscriptions Reach 5 Billion*, iSuppli Report, September 2010.

[2] Gartner, Gartner estimates ICT industry accounts for 2 percent of global CO_2 emissions. <http://www.gartner.com/it/page.jsp?id=503867>, 2007.

[3] Gartner, *Smart 2020: Enabling the Low Carbon Economy in the Information Age*, A Report by The Climate Group, 2008.

[4] H. Cao, *Linearization of High Efficiency Transmitters for Wireless Communications*, PhD Thesis, 2011.

[5] N. S. Networks, *Etsi rrs05 024*, Nokia Siemens Networks Publication, 2011.

[6] M. Spirito et al., Power amplifier PAE and ruggedness optimization by second harmonic control, in *Proceedings of the Bipolar/BiCMOS Circuits and Technology Meeting, 2002*. pp. 173–176, 2002.

[7] N. Sokal and F. Raab, Harmonic output of class-E RF power amplifiers and load coupling network design, *IEEE Journal of Solid-State Circuits*, vol. 12, no. 1, pp. 86–88, February 1977.

[8] Y. Y. Woo, Y. Yang, and B. Kim, Analysis and experiments for high-efficiency class-F and inverse class-F power amplifiers, *IEEE Transactions on Microwave Theory and Techniques*, vol. 54, no. 5, pp. 1969–1974, May 2006.

[9] W. H. Doherty, A new high efficiency power amplifier for modulated waves, *Proceedings of the Institute of Radio Engineers*, vol. 24, no. 9, pp. 1163–1182, 1936.

[10] W. C. Neo et al., A mixed-signal approach towards linear and efficient N-way Doherty amplifiers, *IEEE Transactions in Microwave Theory and Techniques*, vol. 55, no. 5, pp. 866–879, May 2007.

[11] J. Gajadharsing, O. Bosma, and P. van Westen, Analysis and design of a 200 W LDMOS based Doherty amplifier for 3G base-stations, *IEEE MTT-S International Microwave Symposium Digest*, vol. 2, pp. 529–532, June 2004.

[12] M. J. Pelk, W. C. E. Neo, J. R. Gajadharsing, and L. C. N. deVreede, A high-efficiency 100-W GaN three-way Doherty amplifier for base-station applications, *IEEE Transactions in Microwave Theory and Techniques*, vol. 56, no. 7, pp. 1582–1591, 2008.

[13] T. Yamamoto, T. Kitahara, and S. Hiura, 50% drain efficiency Doherty amplifier with optimized power range for W-CDMA signal, in *IEEE MTT-S International Microwave Symposium Digest*, pp. 1263–1266, June 2007.

[14] S. C. Cripps, *RF Power Amplifiers for Wireless Communication*, 2nd ed. Norwood, MA: Artech House Inc, 2006.

[15] N. Ui, H. Sano, and S. Sano, A 80 W 2-stage GaN HEMT Doherty amplifier with 50dBc ACLR, 42% efficiency 32dB gain with DPD for W-CDMA base stations, in *Proceedings of the IEEE Radio Spectrum Conservation Technology Conference*, pp. 1259–1262, June 2007.

[16] D. M. Pozar, Ed., *Microwave Engineering*. John Wiley & Sons, 2005.

[17] J. H. Qureshi et al., A 90-W peak power GaN outphasing amplifier with optimum input signal conditioning, *IEEE Transcations in Microwave Theory and Techniques*, vol. 57, no. 8, pp. 1925–1935, August 2009.

[18] R. M. Fano, Theoretical limitations on the broadband matching of arbitrary impedances, *Journal of the Franklin Institute*, vol. 249, pp. 57–83, 1950.

[19] H. W. Bode, *Network Analysis and Feedback Amplifier Design*. New York: Van Nostrand, 1945.

[20] S. Wood et al., Advances in high power GaN HEMT transistors, *Microwave Engineering Europe*, May 2009.

[21] J. H. Qureshi et al., A wideband 20-W LDMOS Doherty power amplifier, in *IEEE MTT-S International Microwave Symposium Digest*, pp. 1504–1507, May 2010.

[22] H. Chireix, High power outphasing modulation, *Proceedings of the Institute of Road Engineers*, vol. 23, no. 11, pp. 1370–1392, November 1935.

[23] D. Cox, Linear amplification with nonlinear components, *IEEE Transactions on Communications*, vol. 22, no. 12, pp. 1942–1945, December 1974.

[24] I. Hakala et al., A 2.14-GHz Chireix outphasing transmitter, *IEEE Transactions on Microwave Theory and Techniques*, vol. 53, no. 6, pp. 2129–2138, June 2005.

[25] J. H. Qureshi et al., A 90-W peak power GaN outphasing amplifier with optimum input signal conditioning, *IEEE Transactions on Microwave Theory and Techniques*, vol. 57, no. 8, pp. 1925–1935, August 2009.

[26] W. Gerhard and R. Knoechel, Novel transmission line combiner for highly efficient outphasing RF power amplifiers, in *Microwave Integrated Circuit Conference, 2007. EuMIC 2007. European*, pp. 1433–1436, October 2007.

[27] R. Beltran, F. Raab, and A. Velazquez, HF outphasing transmitter using class-E power amplifiers, in *Microwave Symposium Digest, 2009. MTT '09. IEEE MTT-S International*, pp. 757–760, June 2009.

[28] M. P. van der Heijden, M. Acar, J. S. Vromans, and D. A. Calvillo-Cortes, A 19W high-efficiency wide-band CMOS-GaN class-E Chireix RF outphasing power amplifier, in *IEEE MTT-S International*, pp. 1–4, June 2011.

[29] M. Acar, A. Annema, and B. Nauta, Generalized analytical design equations for variable slope class-E power amplifiers, in *13th IEEE International Conference on Electronics, Circuits and Systems, 2006. ICECS '06*, pp. 431–434, December 2006.

[30] D. Calvillo-Cortes et al., A 65nm CMOS pulse-width-controlled driver with 8vpp output voltage for switch-mode RF pas up to 3.6GHz, in *2011 IEEE International Solid-State Circuits Conference Digest of Technical Papers (ISSCC)*, pp. 58–60, February 2011.

[31] J. Sonsky et al., Innovative high voltage transistors for complex HV/RF SoCs in baseline CMOS, in *International Symposium on VLSI Technology, Systems and Applications, 2008. VLSI-TSA 2008.* pp. 115–116, April 2008.

[32] E. Cijvat and H. Sjoland, Two 130nm CMOS class-D RF power amplifiers suitable for polar transmitter architectures, in *9th International Conference on Solid-State and Integrated-Circuit Technology, 2008. ICSICT 2008*, pp. 1380–1383, October 2008.

[33] M. Acar et al., Scalable CMOS power devices with 70 percent PAE and 1, 2 and 3.4 Watt output power at 2GHz, in *Radio Frequency Integrated Circuits Symposium, 2009. RFIC 2009*, IEEE, pp. 233–236, June 2009.

[34] M. Acar, M. P. van der Heijden, and D. M. W. Leenaerts, 0.75 Watt and 5 Watt drivers in standard 65nm CMOS technology for high power RF applications, *Radio Frequency Integrated Circuits Symposium (RFIC)*, pp. 283–286, 2012.

Digital Fractional-N Frequency Synthesis

5

Salvatore Levantino and Carlo Samori

Politecnico di Milano, Italy

INTRODUCTION

One of the significant trends in microelectronics research, started during last decade, is the effort to improve the performance of analog circuits exploiting digital circuits. In most of the situations, digital circuits are employed for the calibration and the correction of analog impairments, while, in some special cases, they may even replace their analog counterparts. The motivation behind this tendency is, at least, twofold. On one side, nanoscaled CMOS technologies enable powerful digital signal processing both at low cost and reduced area occupation. On the other side, achieving high performance in analog circuits often becomes tougher in ultra-scaled technologies. One additional advantage of the digital-assistance of analog circuits is the easy portability of a digital design to a new technology process, which can substantially reduce the design effort and the time-to-market. These benefits are particularly evident in large system-on-chips, where the analog front-ends are integrated with digital processing.

The charge-pump phase-locked loop (PLL), typically employed as a frequency synthesizer, in the radio front-end of wireless systems, is the analog building block that has taken more advantage from this progress than any other. As a matter of fact, PLLs represent a typical example in which analog performance degrades in scaled CMOS processes' For instance, the value of the charge-pump current cannot scale down because it is set by the noise constraint and the quality of the charge-pump current generators degrades as the voltage supply scales down. The area occupation of the loop filter does not scale too and as a result the filter may be very bulky. In some cases, it must be realized with external components and it cannot be easily programmed to change to the PLL bandwidth. Spur-cancellation algorithms such as those based on correlation must be applied to analog variables relying on analog blocks, which increase power consumption and provide limited cancellation. By contrast, digital PLLs eliminate the charge pump ("their loop filter being digital scales with new technology nodes and it is programmable"); digital calibration techniques are naturally implemented with limited area occupation and power consumption.

Manganaro: Advances in Analog and RF IC Design. http://dx.doi.org/10.1016/B978-0-12-398326-8.00005-4

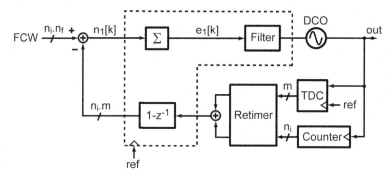

FIGURE 5.1

Phase-domain all-digital PLL.

Digital PLLs for wireless applications, i.e. featuring low noise and low spurs, are usually divided into two categories: The all-digital PLL (ADPLL) also referred to as phase-domain or divider-less PLL and the digital $\Delta\Sigma$ fractional-*N* PLL.

The ADPLL, whose first application to wireless systems was introduced by Texas Instruments in 2004 [1], has the simplified architecture depicted in Figure 5.1. The output of the oscillator drives concurrently a counter and a time-to-digital converter (TDC). The equivalent topology in [1] omits the discrete-time difference and shifts the accumulator out of the loop. The second topology introduced in [2] is the digital $\Delta\Sigma$ fractional-*N* PLL in Figure 5.2. It resembles an analog $\Delta\Sigma$ fractional-*N* PLL [3],

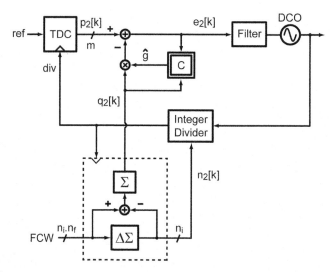

FIGURE 5.2

$\Delta\Sigma$ fractional-*N* digital PLL.

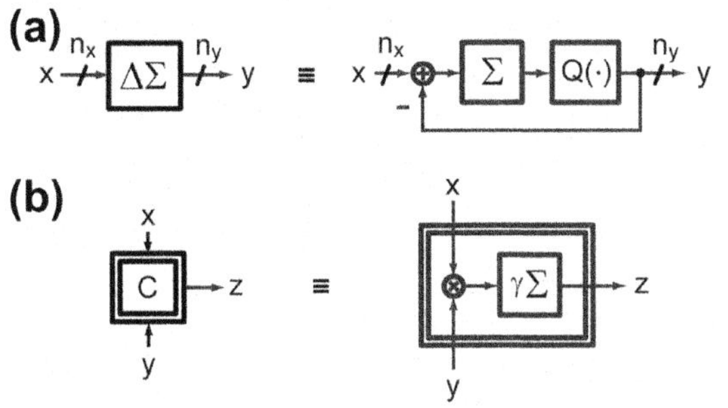

FIGURE 5.3

Description of building blocks: (a) First-order $\Delta\Sigma$ modulator. (b) correlator.

in which the TDC replaces the phase-frequency detector (PFD) and the charge pump. The digital $\Delta\Sigma$ modulator converts the FCW word with (n_i+n_f) bits into a digital word with n_i bits to be fed to the integer-N divider. The added quantization error has the high-pass-shaped spectrum, typical of $\Delta\Sigma$ converters. The block schematic of a first-order digital $\Delta\Sigma$ modulator is shown in Figure 5.3a, where the block $Q(\cdot)$ is a standard quantizer truncating the additional bits. Similarly to analog implementations of fractional-N PLLs, the PLL topology in Figure 5.2 includes a least-mean-square (LMS) scheme for the cancelation of $\Delta\Sigma$ quantization noise and spurs [4]. The correlator C, whose block schematic is shown in Figure 5.3b provides an estimation of the gain of the divider/TDC cascade and practically implements the LMS algorithm.

Both the ADPLL and digital-$\Delta\Sigma$-PLL circuits adopt a digital loop filter and a digitally controlled oscillator (DCO). In wireless applications, the latter is usually implemented as an LC-type oscillator, tuned by switching capacitors of fixed values. In both topologies, the output frequency f_{dco} is equal to the frequency of the reference signal f_{ref} multiplied by the frequency control word (FCW). In general, FCW can be written as $(N+\alpha)$, where N and α are the integer and the fractional parts, respectively. While N is represented by n_i bits, α is represented by n_f bits.

The behavior of the ADPLL and of the digital $\Delta\Sigma$ fractional-N PLL is recalled in the following section. We will show that after the introduction of the LMS cancelation scheme in the second topology, the two structures achieve the same theoretical fractional-spur level induced by the finite TDC resolution.

From Figures 5.1 and 5.2 it is also apparent that the term "digital PLL" is an overstatement. In fact both PLLs belong to the class of mixed analog/digital circuits. The TDC is an analog-to-digital converter [5], while the DCO converts a digital signal (the tuning word) into the frequency of its output periodic signal. The simultaneous presence of these conversions within a feedback loop gives rise to quantization noise and concentrated spurious tones. Both the frequency resolution of the DCO and the time resolution of the TDC improve as technology scales down, since smaller capacitances can be switched on and off and faster delay stages are available respectively. However, for a given technology node, while finer DCO frequency resolution can be achieved relying on oversampling, i.e. interpolating the slow loop-filter output with a faster digital $\Delta\Sigma$ modulator, and on the filtering action of the DCO itself, increasing the time resolution of the TDC is much tougher and it is usually paid with larger power consumption, as in any standard ADC. For this reason, many recent works have focused in particular on improving TDC resolution at low power consumption [6–14]. The subject of TDC-induced noise and spurs becomes even more problematic when two other issues are taken into account, namely the quest for wide PLL bandwidths and the nonlinearity of the TDC. Even if the nature of these issues is different, the former is a design specification while the latter is a circuit nonideality, both contribute to increase noise and spur levels.

The width of the PLL bandwidth is gauged against the reference frequency, f_{ref}, which is typically derived from a quartz oscillator usually operating in the 20–50 MHz range. To guarantee a safe stability margin, the PLL bandwidth cannot exceed a value in the order of 1/10th of its reference frequency, f_{ref}, [15] and it is ordinarily set to a much lower value, i.e. around few tens of kHz. A PLL featuring a bandwidth larger than 100 kHz, up to the limit of some MHz, is therefore considered wideband. Wideband PLLs entail the advantages of (i) speeding up channel switching, (ii) counteracting the pulling phenomena caused by other switching blocks, as for instance power stages integrated on the same die, and (iii) filtering out the close-in $1/f^3$ phase noise of the DCO, which can be the dominant contribution well beyond 1 MHz in ultra-scaled CMOS.

Therefore, a frequency synthesizer based on a wide-bandwidth PLL is useful when the integral-phase-noise requirements are tight, such as in WiMAX radios, since the oscillator specifications can be relaxed, saving power consumption, but becomes necessary in cellular standards, such as LTE, in order to achieve the spot noise performance with a high $1/f^3$ oscillator noise. As a drawback, the larger the bandwidth, the lower the attenuation of the loop filter on noise and spurs induced by the TDC and by other input blocks. Nonlinearity of the TDC merely aggravates the latter issue, producing in-band spurs, which are attenuated by the loop filter and whose frequency is difficult to predict. An improved TDC linearity, like its time resolution, can be achieved by increasing the power dissipation.

For these reasons, this chapter is mainly focused on the different techniques and architectures that have been recently proposed to alleviate the nonlinearity of the TDC at low power and to widen PLL bandwidth. Other techniques for high-resolution TDC design can be found in Chapter 12. We will start with a comparison of the two architectures in Figures 5.1 and 5.2 in term of spurs, first considering an ideally linear TDC and then taking into account the effect of nonlinearity. We will then review the energy-efficient digital techniques, which mitigate TDC nonlinearity. Finally, these considerations will lead us to drastically modify the fractional-*N* digital PLL into a *digital-to-time-converter-based PLL*. In the latter, the multibit TDC is replaced by a digital-to-time converter (DTC) and by a single-bit TDC, which is by definition not affected by nonlinearity and is implemented as a simple thus low-power flip-flop.

5.1 Performance of digital PLLs
5.1.1 Noise–power trade-off and spur performance

The request for RF local oscillators (LO's) with higher spectral purity, featuring low phase noise and spurs, is expected to increase with the advent of the fourth-generation (4G) communications standards, such as LTE and WiMAX. In cellular standards from GSM to LTE, the most stringent requirement derives from the presence of large blockers and from the phenomenon of reciprocal mixing, and it affects the spot phase noise of the LO, i.e. the phase noise at a certain frequency offset. As an example, GSM typically requires an LO with phase noise lower than $-162\,$dBc/Hz at 20 MHz offset. In wireless LAN standards, such as IEEE 802.11 and the later 802.16 (WiMAX), the integral of the LO phase noise spectrum is instead the toughest specification, which may reach a value of $-35\,$dBc or about one degree rms. This requirement can be alternatively expressed in terms of the *absolute jitter*, which is the standard deviation of the difference between the zero-crossing instants of the LO signal with respect to an ideal noiseless clock. For instance, -35 dBc phase noise translates into 800 fs absolute jitter of a 3.6 GHz LO. Both phase noise and spurs affect the integral phase noise, which typically trades with power consumption.

Digital PLLs, especially those used as fractional-N frequency synthesizers, may substantially improve this trade-off, since scaled technologies may enable complex calibration and spur-cancelation algorithms operating at low power. However, we need to bear in mind that the presence of the TDC, which may be one of the most power consuming-blocks, often sets the jitter–power trade-off. Therefore, any digital correction algorithm based on a high-resolution and/or high-linearity TDC could hardly improve the jitter–power trade-off.

In both of the architectures in Figures 5.1 and 5.2, the TDC quantization noise (associated to the finite TDC time resolution) contributes to the in-band phase noise in the same way. For the quantification of this contribution under the assumption of uniform distribution of quantization noise, the reader may refer to Chapter 12. The flat phase noise evaluated in this fashion represents a kind of ultimate limit to the PLL in-band noise and it is usually largely dominated by other contributions. Let us consider the example of a 3.6 GHz PLL with a 40 MHz reference clock and a TDC with 1 ps resolution (i.e. a 9 bit TDC, assuming that its input range must accommodate an entire DCO period with margin). Its quantization noise should produce an in-band noise at the PLL output of about -120 dBc/Hz, a figure that is hardly achieved in digital PLLs. The latter consideration should not lead to the erroneous conclusion that, after all, digital PLLs for wireless do not need very high-resolution TDCs. As demonstrated in [16,17] for the ADPLL architecture, the finite resolution of the TDC instead of producing broadband quantization noise gives rise to spurious tones typically falling within PLL bandwidth (when near-integer channels are synthesized) and substantially increasing output jitter. The problem becomes even more serious in the case of wide-bandwidth PLLs and when accounting for TDC nonlinearity. The same conclusions on the effects of TDC finite resolution and nonlinearity can be drawn for the alternative topology in Figure 5.2 [18]. In the following subsections, we will review these concepts.

A lower boundary for the achievable level of spurs is indeed set by supply disturbances and parasitic signal coupling. The interaction between stages switching at the fractional frequency (such as the integer-*N* divider) and the DCO typically produces unwanted frequency modulations of the carrier at the fractional frequency. These phenomena, taking place as in any mixed analog/digital circuits, are very difficult to quantify and demand special care in reducing the sensitivity of the victim and the block-to-block coupling.

5.1.2 TDC finite resolution in the ADPLL topology

In the ADPLL topology in Figure 5.1, the output of the counter accumulates the number of integer periods of the DCO output signal. Therefore, it provides a coarse measurement of the output phase of the DCO. Let us assume, for the moment, that the PLL must synthesize an output frequency, $f_{dco} = 1/T_{dco}$, which is an integer multiple N of the clock reference frequency, $f_{ref} = 1/T_{ref}$, meaning that the frequency control word is an integer number: $FCW = N = f_{dco}/f_{ref}$. The finite difference $(1 - z^{-1})$, clocked at f_{ref}, provides the number of the DCO periods within T_{ref} and this digital information is subtracted from FCW to provide the error signal. In this scheme, the frequencies of the signals (and not their phase) are compared, since the output phase is differentiated.

Instead, if a fractional ratio f_{dco}/f_{ref} is desired, the number of DCO periods within T_{ref} is not integer, but the remainder of the division is a fraction of the DCO period equal to αT_{dco}. In this case, the TDC comes into play. It measures the fractional time error between the output and the reference signals (that is proportional to the phase error) with a resolution of m bits. Since the maximum error to be measured is T_{dco}, the TDC must accommodate with some margin this input range. A retiming block, whose design may present some difficulties [19,20], is typically used to combine the information coming from the asynchronous counter and that coming from the TDC clocked at T_{ref}.

The problem with the fractional channel arises because the number of fractional bits of FCW dictated by the desired frequency resolution is usually much higher than the number of bits of the TDC, limited by practical reasons ($n_f > m$). Therefore, if a channel with resolution finer than that allowed by the TDC must be synthesized, the output frequency can converge to the wanted one (FCW times f_{ref}) only on average. Figure 5.4 shows what happens in the simple case in which the PLL is locked and

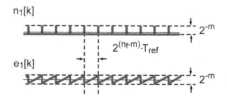

FIGURE 5.4

Typical waveforms of the phase-domain ADPLL.

the FCW has only its LSB equal to one among its last (n_f-m) bits, i.e. the output frequency is close to an integer channel. For the sake of simplicity, starting from the synchronous rising edges of the reference and the DCO signals, we see that their time shift increases to $2^{-nf}T_{dco}$ after one reference cycle. Since the TDC cannot resolve it, this time error piles up in the following cycles and reaches one LSB of the TDC only after $2^{(nf-m)}$ reference cycles.

The error signal $n_1[k]$ in Figure 5.1 will obviously be affected by a quantization error $n_1[k]$, whose average must be zero, given the presence of the subsequent accumulator. The time diagram of $n_1[k]$ is sketched in Figure 5.4, where the time duration of each pulse is T_{ref} and the value of the time error corresponding to $n_1[k]$ is labeled on the right-hand side of the plots. The sequence $n_1[k]$ at the input of the accumulator produces the sawtooth signal $e_1[k]$ at the filter input, whose output modulates the DCO tuning and it is thus responsible for fractional spurs at a frequency offset $f_{ref}/2^{(nf-m)}$ from the carrier. The straightforward solution to this issue is narrowing the loop filter below this frequency offset. However, the PLL bandwidth would be narrower than the desired frequency resolution. This is unacceptable in many practical cases.

5.1.3 TDC finite resolution in the $\Delta\Sigma$-fractional-N topology

We consider now the $\Delta\Sigma$ fractional-N PLL in Figure 5.2, assuming that the first-order $\Delta\Sigma$ modulator depicted in Figure 5.3 is employed and that the resolutions of the FCW word and of the TDC converter are the same as in the ADPLL considered before. The $\Delta\Sigma$ output, $n_2[k]$, increments the division factor by one (corresponding to a time increment of T_{dco}), for a duration of one reference period, every $2^{nf}T_{ref}$ seconds (see the diagram of n_2 in Figure 5.5). In the general case of FCW $=(N+\alpha)$ with integer N and rational α, this period will be T_{ref}/α. The resulting phase error

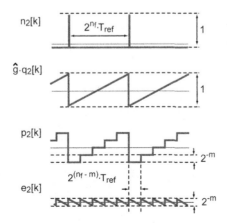

FIGURE 5.5

Typical waveforms of the $\Delta\Sigma$ fractional-N digital PLL.

ramp is much larger than in the ADPLL case. But, since this ramp is proportional to the accumulated $\Delta\Sigma$ quantization error $q_2[k]$, it may be canceled out at the TDC output. To obtain a perfect cancelation, $q_2[k]$ has to be multiplied by the exact TDC gain. An estimation, $\hat{g}$, of this gain is obtained by correlating $q_2[k]$ and the m-bits TDC output $p_2[k]$, employing the typical LMS algorithm for noise canceling [2]. This operation is carried out by the correlation block C, whose block schematic is depicted in Figure 5.3b. In practice, the correlation between the error signal, $e_2[k]$, and the quantization error, $q_2[k]$, is calculated. At steady state, the output of the block C will remain constant on average, which means that the inputs $e_2[k]$ and $q_2[k]$ are uncorrelated and the quantization noise is canceled out. This is possible only if $\hat{g}$ is equal on average to the TDC gain. The higher the constant γ of the correlator C, the faster is the convergence of the LMS loop, but the larger is the fluctuation of $\hat{g}$ around its average value, which may increment PLL in-band noise.

Unfortunately, as in the previous PLL topology, the TDC time resolution $(2^{-m}T_{dco})$ limits the cancelation accuracy. In fact, it implies that the number of TDC bits is much less than the number of bits of the signal subtracted from the TDC output and this gives rise to the staircase $p_2[k]$ shown in Figure 5.5. The time diagram of the TDC output after correction, $e_2[k]$, is identical to the signal $-e_1[k]$ obtained for the ADPLL topology and hence it produces the same fractional spur in the output spectrum. It is possible to verify that the situation does not change employing higher-order $\Delta\Sigma$ modulators in the digital-$\Delta\Sigma$ PLL topology, since the limited TDC resolution still produces the same effects.

The first harmonic of the resulting spurious tone falls at $f_{ref}/2^{(nf-m)}$ in both cases, which is a low value, taking into account that the TDC is limited to about $m = 10$ bits by practical constrains and that $n_f = 20$ bits of fractional resolution may required to get 40 Hz resolution from a 40 MHz reference signal. Thus, the resulting tone will fall in the worst case of a near-integer channel at 40 kHz, which in many cases will fall within the loop bandwidth. This simple analysis suggests that in both the PLL architectures, the fractional spur can be reduced by increasing TDC resolution; therefore, in practice, by increasing the number of TDC bits.

In the previous analysis, we neglected the presence of the thermal and flicker noise sources associated with the various PLL building blocks (such as reference oscillator, TDC, and DCO). If the TDC quantization noise is comparable or lower than random noise, the error ramps in Figures 5.4 and 5.5 are dithered and the resulting fractional-spur level in the output spectrum would be reduced with respect to its theoretical value (see for instance Figure 12.10). Additional random noise or pseudo-random dithering sequence can also be artificially injected into the loop in order to realize such an effect; however, the level of fractional spurs is reduced at the cost of higher phase noise.

Moreover, in the analysis, the TDC is assumed to be ideally linear. Therefore, if its resolution was infinite, no spurs would appear at the output even when a fractional channel is synthesized (i.e. when α is nonzero, being FCW $= N + \alpha$). This is because the difference between the sampled phases of DCO and reference varies at a frequency equal to $\alpha \cdot f_{ref}$, but a TDC with infinite resolution would be able to measure this difference and to provide the correct feedback signal. Unfortunately, the presence of TDC nonlinearity not only causes higher level of the fractional spurs but it also produces unexpected spectrum folding of the random dithering signals injected to reduce them and makes this solution often ineffective.

5.1.4 **Effects of TDC nonlinearity**

The effects of the TDC nonlinearity in the ADPLL topology of Figure 5.1 were discussed in [16]. It is clear that a nonlinear conversion characteristic, as in any ADC, would produce spurious tones in its output spectrum. In a PLL, this ultimately produces spurs at the output, whose level increases with the integral nonlinearity (INL) of the TDC characteristic.

Concerning the second PLL topology in Figure 5.2, the equivalence between the two architectures holds also from the standpoint of TDC nonlinearity, as highlighted in Figure 5.6. The TDC output ramp (represented now as a continuous line, for the sake of simplicity) is distorted and cannot be completely canceled out after subtraction of the estimated linear ramp $\hat{g} \bullet q[k]$. The resulting error $e[k]$ is periodic and it will produce residual fractional spurs at the output [18].

In general, it is very difficult but not to accurately calculate the level of in-band spurs, taking into account the coexisting effects of the finite TDC and DCO resolution [21,22], the presence of random noise sources in the loop and the nonlinearity of TDC. Circuit simulations become, more than ever, an essential tool. To gain

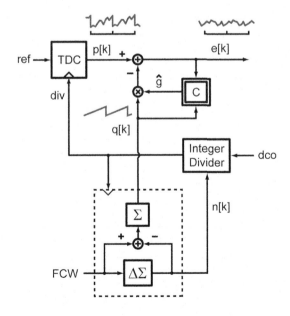

FIGURE 5.6

Effect of TDC nonlinearity.

an intuitive feeling, we can refer to the results presented in [18] and concerning a 3.6 GHz frequency synthesizer with 3 MHz bandwidth and 40 MHz reference. Simulations that take into account all the previously cited nonidealities show that spurs are always below -50 dBc even in the worst fractional-*N* channel, if the TDC has resolution of 3 ps and integral nonlinearity (INL) of 3 ps (i.e. ± 0.5 LSB). This requirement is equivalent to a 40 dB dynamic range and INL, relative to a full scale of one output period.

5.2 Techniques for TDC nonlinearity mitigation

The number of works focusing on TDC nonlinearity appears relatively small when compared to the wealth of literature devoted to the increase of its time resolution. Nonetheless, thanks to the interest for digital PLLs, several techniques for nonlinearity mitigation of TDCs have been recently proposed. Oversampled TDCs, such as those based on the concept of the gated ring oscillator (GRO) [8], feature better intrinsic linearity than standard Nyquist-rate converters. For details on this type of converters, the reader may refer to Chapter 12. Other methods are based on an on-chip code density test of the TDC and a subsequent digital correction [16,23,24]. Unfortunately, such methods result in high computational cost and slow convergence. Two other significant examples of linearization, those based on element randomization [18] and large-scale dithering [25], will be discussed in this section.

5.2.1 Element randomization

The mismatches among elements in the classical delay-line TDC represent one main source of TDC nonlinearity [26], identically to what happens in a flash-type ADC. Since the periodicity of the $e[k]$ sequence in Figure 5.6 is responsible for concentrated in-band tones in the spectrum, some kind of randomization can spread the spur energy over the spectrum at the cost of a slight increase in in-band noise. The basic idea behind this approach is depicted in Figure 5.7. The elements setting the time thresholds of the TDC are randomized, adopting an algorithm similar to the dynamic

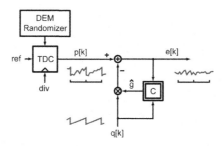

FIGURE 5.7

Element randomization of the TDC.

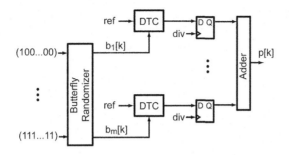

FIGURE 5.8

Implementation of an m-bit flash-type TDC with element randomization.

element matching (DEM) typical of DACs. In this way, since the periodicity of $e[k]$ is destroyed, the mismatches are scrambled and converted into noise, while concentrated tones are avoided. In other words, the mismatches are averaged by the PLL, since even in wideband PLLs its bandwidth is much lower than the scrambling rate (i.e. the reference clock frequency f_{ref}).

While randomization of the elements seems problematic in a delay-line TDC, things change if an m-bit TDC with a parallel topology such as the one sketched in Figure 5.8 is employed. This circuit is realized as an array of $M = 2^m - 1$ digital-to-time converters (DTCs), in parallel by the reference signal, ref. Then, an array of M time arbiters samples the output of the DTCs by means of the feedback signal, div. A DTC can be implemented as the simple delay stage in Figure 5.9. In practice, it can be a digital buffer, either in CMOS or current-mode logic (CML), whose capacitive load is controlled by an r-bit digital word. In this fashion, the input-to-output delay of a DTC can be digitally controlled.

To explain the operation of the circuit in Figure 5.8, let us assume ideally no mismatch among the elements and let us set the M DTCs such that their capacitive loads scale linearly from C to MC. Doing so, the TDC topology is analogous to a voltage-mode flash ADC. The nth time delay is the sum of the fixed intrinsic delay, τ_0, of the stage (hundreds of ps in today's technologies, considering the variable capacitive

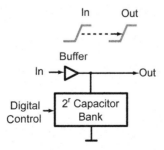

FIGURE 5.9

Implementation of an r-bit digital-to-time converter (DTC).

load that it drives) plus a term, $n\Delta\tau$, depending on the variable load, nC. Because the smallest delay difference, $\Delta\tau$, realizable today is in the order of few ps, it is possible to reach a good time resolution. The thermometric output of the time arbiters could be simply fed to an encoder. The presence of the adder in Figure 5.8, in place of the encoder, is related to the randomization technique we are going to describe.

This circuit can perform element randomization by varying in a pseudo-random fashion the capacitive load connected to the different rows [27]; in other words, the nth time threshold of the TDC is synthesized alternatively by any of the M rows and not always by the nth one.

To analyze this technique in more depth, we have to notice that, being $\tau_0 \gg \Delta\tau$, the dominating source of mismatch is that affecting the intrinsic delay, τ_0. This consideration justifies the presence of the adder in Figure 5.8: let us assume that all the rows are not loaded by the capacitors. The M delays are distributed as a Gaussian probability density function (PDF) around the nominal (average) value τ_0, as shown in Figure 5.10a (for simplicity, we have assumed $M=4$). Thus, thanks to the adder, the output characteristic of the TDC is approximately interpolated by the cumulative distribution function (CDF) of the Gaussian. The more delay stages are connected in parallel, the more the samples of the Gaussian CDF. This characteristic is

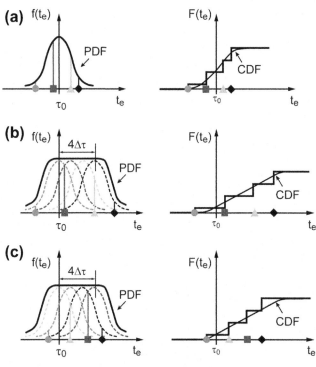

FIGURE 5.10

Threshold distribution and characteristic of a stochastic TDC.

approximately linear around τ_0, even if its input range is very narrow. A TDC operating in this fashion is typically referred to as a *stochastic TDC* [28].

As schematically shown in Figure 5.10b, the use of the M different capacitive loads in each delay stage increases on one side the input linear range of the TDC, since, from the point of view of probability theory, the PDF of the delay of the nth row is the original Gaussian function shifted by $n\Delta\tau$. On the other side, the DEM shuffling the delays among the rows distributes the thresholds on average more uniformly (thus more linearly). The TDC characteristic in Figure 5.10c, which is different from the one is Figure 5.10b, exemplifies the effect of delay shuffling. The effective conversion characteristic of the TDC, given by the average of the many characteristics produced by DEM, is linear around its center.

The 3 MHz-bandwidth $\Delta\Sigma$ fractional-N PLL in [18], adopting such a type of randomization in the TDC reports a worst-case in-band spur of about -57 dBc. The TDC used in that example had 3 ps resolution but only 4 bits ($M=15$). If a larger number of bits were required, the capacitance routing of this TDC scheme would become very consuming in terms of silicon area.

5.2.2 Large-scale dithering

Another approach to the mitigation of TDC nonlinearity consists of injecting a dithering signal in the PLL loop. The beneficial effect of random noise on reducing the level of fractional spurs induced by TDC finite resolution suggests to exploit the same dithering technique for the mitigation of TDC nonlinearity. However, while the effects of TDC quantization can be scrambled by a random signal with a standard deviation comparable to the TDC time resolution, the nonlinearity can be attenuated only with a much larger dithering signal. In practice, the mismatches among elements are effectively averaged only if the entire TDC dynamic range is exploited by the dithering signal. For this reason, such an approach is sometimes called *large-scale dithering*, in contrast to the *small-scale dithering* [29] that it is only effective in counteracting TDC quantization. The large-scale dithering has been proposed and implemented for the linearization of the ADPLL in [25].

Although the injection of a large pseudo-random signal at the TDC input reduces the spur level, it increases the noise level at the TDC output and consequently the in-band noise at the PLL output. It is interesting to note that, from this standpoint, the randomization technique of the TDC elements discussed above can be considered very similar to large-scale dithering. However, as shown in Figure 5.11, the input pseudo-random sequence fed at the TDC input by means of a DTC is known and it can be subtracted at the TDC output. Analogous approaches, referred to as *subtractive dithering*, have been developed in the context of ADC linearization [30–32], while the *sliding-scale method* [33] can be considered a precursor of these techniques.

The dithering sequence, $\delta[k]$, generated by a pseudo-random numbers generator (PRNG) as a digital signal can assume only a finite number of levels. Let us denote as δ_i the generic value of $\delta[k]$ that adds a delay $\delta_i T_{\text{dco}}$. If the pseudo-random delay sweeps entirely the TDC input range, all the delay elements of the TDC will be exerted and the

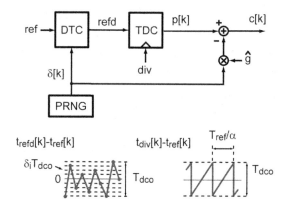

FIGURE 5.11

TDC with injection and first-order cancelation of a large-scale dithering signal.

nonlinear characteristic will be averaged by the PLL loop. In Figure 5.11, the typical evolution of the zero-crossing instants of the div signal $t_{div}[k]$ (in the case of a first-order $\Delta\Sigma$ modulator) is compared to that of the time-dithered reference signal $t_{refd}[k]$.

In order to subtract the dithering sequence from the TDC output, the sequence, $\delta[k]$, must first be multiplied by the gain, $\hat{g}$, of the DTC and the TDC cascade (to be estimated in some fashion). The TDC is linearized by this method, but the DTC may be in general nonlinear. Thus, $\hat{g}$ provides only a linear fitting of its characteristic and only a first-order cancelation of the dither signal can be obtained.

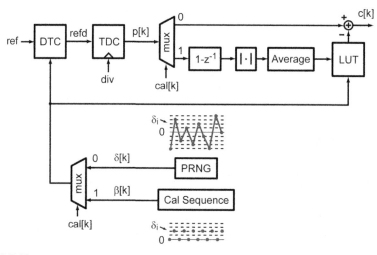

FIGURE 5.12

TDC with large-scale dithering and DTC nonlinearity correction.

The calibration proposed in [25] is slightly more complex because it also faces the issue of DTC nonlinearity and the automatic estimation of its characteristic. It can be simplified as shown in Figure 5.12. In calibration mode, a fractional-N channel is synthesized such that FCW is $(N+\alpha)$ and that the entire input of the TDC is swept by a ramp having slope α. At the same time, in order to calibrate the ith dithering delay, δ_i, we switch on and off the delay, δ_i, alternately at every other TDC sample. Doing so, the magnitude of the difference between two consecutive samples at the output of the $(1-z^{-1})$ block switches between $|\delta_i+\alpha|$ and $|\delta_i-\alpha|$. Thus, the average of an even number of these samples provides an estimate of δ_i.

Thanks to this approach, the 3.5 GHz ADPLL in [25] features a 3.4 MHz bandwidth (from a 35 MHz reference) and achieves a worst-case in-band spur of $-58\,$dBc. This dithering technique may become rather costly in terms of calibration time. In the cited implementation, the DTC has 16 delays and, for each of them, the average is performed over 2^{18} samples in order to obtain 23 bit accuracy. The resulting calibration lasts about $120\,$ms, which must be accommodated in the startup sequence.

5.3 **Digital-to-time-converter-based digital PLLs**

Alternative approaches for relaxing the nonlinearity requirements of the TDC consist of completely rethinking the digital PLL topology. Those schemes relying on DTC-based fractional-N dividers in place of the integer-N ones represent one of the promising evolutions of digital PLLs in this sense. They allow employing TDCs with a much narrower range or ultimately with a single bit. An alternative approach to allow a narrow-range TDC consists of the combination of a digital loop in parallel with an analog loop [34]. Although the concept of realizing a true fractional-N divider was well known [35–37], the application to conventional analog fractional-N PLLs has been prevented by the inaccuracy of these blocks and their poor spectral purity. As will be clearer in this section, the digital PLL makes it possible and relatively easy to assist those dividers and to improve their accuracy.

5.3.1 **Fractional-N digital PLLs with narrow-range TDCs**

The conventional way of synthesizing a fractional-N frequency is toggling the modulus control of an integer-N divider. As we know, if the modulus is increased from N to $(N+1)$, a delay of one T_{dco} is added to the feedback signal of the loop and this delay must be correctly converted by the TDC without clipping, in order to be accurately canceled out at the TDC output. As already discussed in the previous sections, the large spur and noise introduced by the modulus toggling can be eliminated using either a narrow PLL bandwidth or a high-resolution TDC.

However, this problem can be relaxed by designing a true-fractional-N divider, which is able to perform a fractional-N division without toggling its modulus control. This function requires the availability of a DTC, which generates intermediate

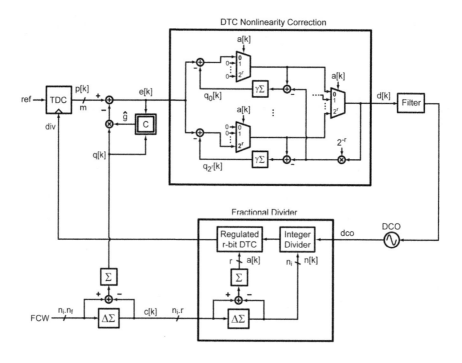

FIGURE 5.13

DTC-based PLL with narrow-range TDC and DTC nonlinearity correction.

delays between 0 and one T_{dco}. Since the DTC dynamic range must be exactly equal to T_{dco}, this block is typically referred to as *digitally-controlled phase interpolator* or *regulated DTC*. If this type of fractional-*N* divider were added into the digital PLL, as shown in Figure 5.13, no quantization would be introduced when synthesizing a fractional-*N* frequency. Obviously, the limited number of bits of the DTC (let us say *r* bit) would give a limited resolution of the frequency synthesizer. Thus, we can increase frequency resolution by toggling the modulus of the fractional-*N* divider. The advantage over the case of the integer-*N* divider is the reduced range of the time error at the TDC input. In particular, the latter is reduced from T_{dco} to the value of the DTC time resolution, or equivalently by a factor of 2^r. For this reason, although the TDC still needs a tight time resolution to guarantee a low level of fractional spurs after the quantization-error cancelation, it can have a reduced dynamic range. In practice, a TDC with a much lower number of bits can be employed without causing signal clipping. This modification results in two main advantages. First, the TDC systematic nonlinearity is intrinsically lower since the dynamic range is narrower. Second, a reduced number of bits enables a flash-type implementation of the TDC with a small number of elements; thus, a linearization method which counteracts the effects of element mismatches, such as that based on the element randomizer, can be easily and practically employed.

The limited input linear range of the TDC could considerably slow down the PLL frequency-lock transient. However, adding a secondary loop based on a coarse TDC with wider dynamic range easily solves this issue.

A possible implementation of the r-bit DTC is shown in Figure 5.14 [38]. It consists of a delay line with variable delay stages and a multiplexer selecting one of the generated phases. The dynamic range of the delay line is automatically adapted to be exactly equal to the DCO period T_{dco} or to a multiple of T_{dco} by varying the delay of the stages. In practice, this scheme creates a delay-locked loop (DLL). A first D-type flip-flop creates P_{0d} delayed by T_{dco} from P_0. Then, a second flip-flop detects if P_{0d} leads or lags the last phase, P_{2r}, acting in practice as a lead-lag phase detector. The single output bit of this detector is accumulated by a digital integrator, then fed to a first-order digital $\Delta\Sigma$ modulator, filtered by a passive RC filter and finally applied to the analog control of the stage delays. At steady state, the loop aligns P_{0d} and P_{2r}, as shown in the signal diagrams in Figure 5.15; thus, it creates a delay of one T_{dco} between P_0 and P_{2r} and a delay of $T_{dco}/2^r$ between each two subsequent signals P_i and P_{i+1} (in the absence of mismatches among the stages). The $\Delta\Sigma$ modulator and the passive filter allow the avoidance of the limit cycles typical of digital DLLs, which would degrade the deterministic jitter of the synthesized phases. In practice, deterministic jitter lower than tens of fs are achievable with $r=4$ and with a factor of about 100 between the $\Delta\Sigma$ clock and the DLL bandwidth.

Alternative implementations of phase interpolators consist of multiphase DCO topologies (such as coupled LC oscillators or ring oscillators [39]), injection-locked multiphase oscillators or divider circuits [40].

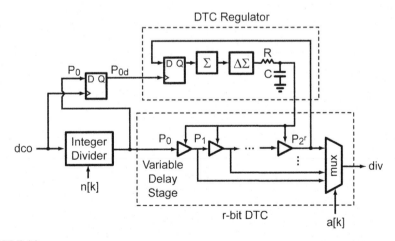

FIGURE 5.14

Implementation of the regulated r-bit DTC.

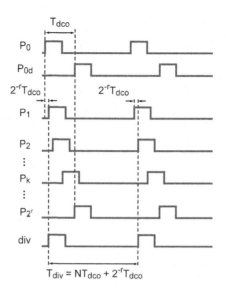

FIGURE 5.15

Typical waveforms in the regulated *r*-bit DTC.

Unfortunately, the straightforward application of a phase-interpolator-based fractional-*N* divider in a PLL would suffer from the nonlinearity of the DTC arising from the mismatches of its elements. Apparently, introducing the DTC in the digital PLL has just shifted the nonlinearity issue from the TDC to the DTC, which would represent an additional source of unwanted spurs. Actually, the situation has radically changed for three main reasons: (i) the required number of bits in the DTC is as low as the number of bits required in the TDC (i.e. a PLL with an *m*-bit TDC and *r*-bit DTC has performance comparable to a PLL with an $(m+r)$-bit TDC); thus, the impact of element mismatches in both converters can be reduced by using larger devices with limited total area occupation. (ii) The DTC suffers less from element mismatches if compared to a TDC, since in the latter case the flip-flop mismatch is an additional contribution to nonlinearity. (iii) The nonlinearity of the DTC can be corrected by relying on a background digital algorithm, because we can detect the error produced at the TDC output for each input of the DTC, correlate this error with the input of the DTC and correct for it. The latter algorithm will be illustrated in the following subsection.

5.3.2 Digital background correction of DTC nonlinearity

In the previously discussed topology of regulated DTC, the inaccuracies among the generated phases P_n arise for instance from mismatches among the delay cells. If one of the delay elements is not matched to the others, the total delay of the line is still forced to be equal to T_{dco} by the DLL loop, but it is no more uniformly distributed among the

2^r elements. Another source of inaccuracy is the delay mismatch of the gates of the multiplexer. Both of those effects produce periodic modulations of the zero crossings of the divider output, which may cause non-negligible spurs in the PLL spectrum. Assuming the presence of a delay error, t_k, associated to the kth phase, P_k, the divider output would periodically contain this shift, every time the phase P_k is selected. This periodic modulation would obviously appear at the TDC output.

Methods aiming to remove spurs caused by DTC nonlinearity in phase-interpolator-based dividers typically try to detect the phase error and to tune the position of each phase in order to cancel out the error. However, the adoption of those methods requiring several feedback loops in the analog domain complicates the design and typically increases power dissipation. Other methods based on dithering the phase selection signal have the advantage of the digital implementation, but suffer from limited attenuation of spurious tones and addition of high-frequency noise.

In principle, instead of correcting the analog position of signal edges, in the specific case of a PLL, it is sufficient to cancel out the error associated with the DTC phase mismatches from the output of the TDC and not necessarily from the DTC itself. Doing so, the spur cancelation can be entirely implemented in the digital domain with no noise penalty and negligible power dissipation. Since the TDC is able to detect the phase error of the phase interpolator, we take advantage of the correlation existing between the TDC output, $p[k]$, and the phase selection signal, $a[k]$.

This cancelation can be performed by means of the bank of 2^r time-variant digital filters connected to the TDC output and shown in Figure 5.13 [18]. The kth filter contains an integrator with gain γ, which accumulates the digitized time error, t_k, of the kth phase, P_k. When $a[k]$ selects the nth phase of the phase interpolator, the resulting $p[k]$ is routed to the nth filter. In this fashion, the 2^r accumulators closed in feedback will converge to the 2^r mismatch errors of the delay line (assuming $\gamma < 1$). An extended analysis of this filter bank can be found in [41]. The $\{q_n\}$ vector which stores at steady state the corrections for the 2^r phases of the delay line is subtracted from $p[k]$. Thanks to the averaging action of the filters, the error in the estimation of mismatches can be much lower than the TDC quantization noise.

To avoid canceling a constant $p[k]$ sequence, which would add a zero at dc in the PLL loop gain and would make the PLL a type-I loop, the output, $e[k]$, is fed back to the input of each integrator after multiplication by 2^{-r}. In order to verify the response of the filter bank at dc, let us suppose that the constant $p[k] = h$ is not filtered out and it appears at the output $e[k]$. Thanks to the subtraction of $2^{-r}h$ from $e[k]$, the input of the 2^r integrators has zero average. Consequently, their outputs are zero plus a ripple, that is negligible if $\gamma \ll 1$, and the sequence $e[k]$ is equal to h as initially assumed.

Assuming the sequence $a[k]$ to be a periodic sawtooth waveform, as obtained from the digital accumulator with constant input, this correction system can be regarded as a multirate filter and its frequency response can be defined. The response is obtained by calculating the FFT of $e[k]$ after (i) applying a periodic sawtooth sequence, $a[k]$, and (ii) superimposing a random white noise to $p[k]$ (that models the practical noise of the PLL building blocks). The resulting frequency response in Figure 5.16 shows notches at multiples of $f_{\text{ref}}/2^r$, which demonstrates the desired

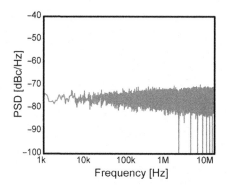

FIGURE 5.16

Transfer function of the DTC-nonlinearity-correction algorithm.

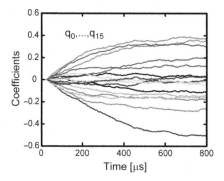

FIGURE 5.17

Coefficient convergence of the DTC-nonlinearity-correction algorithm.

property of canceling out the spurs originating from mismatches. The algorithm is effective even when $a[k]$ is not exactly a sawtooth, as when it is obtained from a $\Delta\Sigma$ modulator with nonzero input. In that case, the fractional spur is slightly offset from $f_{ref}/2^r$, but it is still canceled out. The convergence of the correction coefficients $\{q_n\}$ resulting from simulation of the close-loop PLL with mismatched delay line ($r=4$) and filter bank with $\gamma=2^{-10}$ is shown in Figure 5.17. The coefficients settle in about 10^4 clock cycles, while the residual noise of the coefficients is caused by the random noise sources present in the loop.

Compared to foreground calibration methods suffering from lack of tracking capability, this algorithm operates in background mode, performing both estimation and cancelation of mismatch-induced errors during the normal PLL operation and following the drifts of temperature and voltage supply. The implementation of this algorithm, although based on correlation, does not require any multiplication, as simple multiplexers implement the correlators. This results in reduced silicon area and power consumption.

The test chip in Figure 5.18 implements in 65 nm CMOS a digital $\Delta\Sigma$ fractional-N PLL [18] with a narrow-range TDC and with the so-far-discussed scheme for the correction of DTC nonlinearity. Measurements confirm that this approach is effective in reducing the in-band fractional spurs induced by DTC nonlinearity from –44 to below –68 dBc. The measured spectra obtained for a near-integer channel, gradually enabling the spur-correction algorithms are shown in Figure 5.19: (a) no correction, (b) cancelation of $\Delta\Sigma$ quantization error, (c) TDC element randomization, (d) DTC nonlinearity correction. The combination of these techniques demonstrates a reduction of all spurs below $-57\,$dBc.

5.3.3 Fractional-*N* digital PLLs with single-bit TDCs

Following the argument discussed in the previous subsections, the deleterious effect of TDC mismatches and systematic errors on fractional spur generation could be in principle removed by adopting a single-bit TDC. A single-bit TDC, which can be implemented in practice as a D-type flip-flop, detects whether the flip-flop D input leads or lags the clock input. In practice, it acts as a hard limiter on the time delay between its two inputs. For this reason, in the field of control theory, this detector is often referred to as *lead–lag detector* or *bang-bang detector*. Given its very simple implementation, the additional advantage of a single-bit TDC over a multi-bit one is obviously the reduction of the synthesizer power consumption, especially considering that the TDC is typically one of the most power-hungry blocks in a digital PLL.

Although widely used in PLLs designed for clock-and-data-recovery applications, thanks to its high speed, this kind of detector is rarely used in the field of frequency synthesizers. Except for the case of integer-N frequency synthesis (where bang-bang detectors have recently been employed both in analog and digital implementations

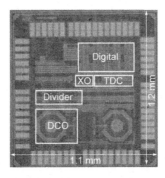

FIGURE 5.18

Die photograph of the digital PLL with narrow-range TDC in [18].

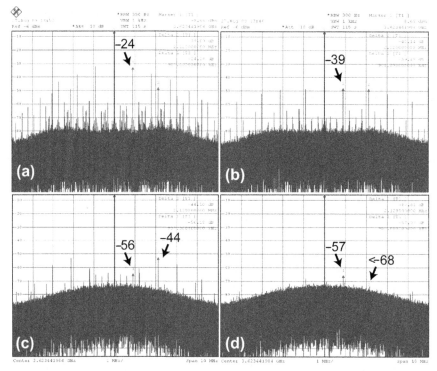

FIGURE 5.19

Measured spectra of the digital PLL in [18] obtained for a near-integer channel, gradually enabling the spur-correction algorithms.

[42]), a nonlinear phase detector is typically avoided. The main reason is that in fractional-*N* PLLs, the $\Delta\Sigma$ dithering of the divider modulus produces high-frequency quantization noise, which is supposed to be filtered out by the loop filter. A highly nonlinear block in the loop such as a bang-bang phase detector produces spectrum folding of this quantization noise and it results in much higher PLL in-band noise.

A possible solution to the nonlinearity of the single-bit TDC consists of closing it in a loop with the $\Delta\Sigma$ modulator [43]. The resulting circuit is the Copeland's frequency discriminator. Since this circuit acts as a frequency detector, the resulting PLL is a type-I, which intrinsically offers poorer phase noise performance. An alternative approach providing noise shaping of the quantization noise of the single-bit TDC can be found in [44].

The issue of dithering could in principle be solved by adopting a DTC-based fractional-*N* divider similarly to the previously discussed case of the low-range TDC. However, since the residual dithering of the PLL time error is as large as the DTC time resolution, the TDC range in the previous case was designed to be greater than or equal to the DTC resolution and the cancelation of the dithering signal was performed at the TDC output. In this case, the TDC would have apparently no linear range and the dithering signal cannot be canceled at the single-bit output of the TDC; both considerations would suggest that the DTC resolution in this case should be infinite.

In practice, thanks to the presence of random noise at the input of the single-bit TDC (such as that caused by flicker- and thermal-induced noise of the PLL building blocks), the DTC does not need an infinite resolution. Good spur and quantization noise cancelation can be obtained by employing a DTC with a time resolution slightly better than the standard deviation of the time error at the input of the TDC and by performing the cancelation of the quantization error at the input of the DTC, instead of applying it at the output of the TDC.

It is worth noting that while a pseudo-random dithering sequence injected into the loop could potentially scramble the spurious tones at the price of higher phase noise, in this case the random physical noise already present in the loop and induced for instance by the reference or the DCO phase noise eliminates the fractional periodicity. The idea that the inevitable presence of physical noise can be useful in the presence of nonlinearities is known in some other scientific disciplines as *stochastic resonance* [45].

In the presence of a dominating random noise at the input of the single-bit TDC, the detector acts as a linear stage and it can be modeled in the phase domain as a constant gain plus quantization noise [46–48]. To understand this behavior, we should note that the PLL with a bang-bang detector and a simple gain stage as loop filter has a phase-domain model made of an integrator (i.e. the DCO) closed in a feedback loop by a hard limiter (i.e. the single-bit TDC). This model is identical to that of a Δ modulator. Thus, the output of the TDC would be given by the Δ-modulation of the reference phase. In practice, if quantization noise is reduced below random noise, good linearity can be achieved even with a hard limiter.

In general, assuming that the random noise at the TDC input has Gaussian distribution with standard deviation σ and that the PLL bandwidth is much lower than the reference frequency, it holds that the average of the single-bit TDC output is proportional to the integral of the Gaussian density of the random noise. The resulting characteristic of the detector (average output vs. input time error) is an error function, which has an approximated linear range as wide as 2σ around zero and a linear gain inversely proportional to σ [46]. The dependence of the PLL loop gain and thus of the PLL bandwidth on the amount of noise σ requires some kind of automatic regulation to guarantee reproducible performance over PVT variations. Moreover, the narrow linear range requires a coarse auxiliary TDC and an additional control loop to speed up PLL lock transient. Methods for the solution of both issues can be found in [49].

If the reference jitter is much lower than the PLL output jitter, the latter dominates the random noise at the TDC input. Thus, typical values of random time error σ in high-frequency frequency synthesizers may be of the order of 1 ps or less. Thus, since DTC resolution must be slightly lower than random noise and since the range of the DTC must be exactly equal to one T_{dco} (i.e. in the order of 1 ns in the 1-GHz range) as already discussed in the previous subsection, a DTC with at least 10 bits is necessary in practice.

The multibit TDC can be replaced by a DTC even in the phase-domain ADPLL topology by adopting a mid-rise quantization [50]. In this way, the DTC realizes the one-bit phase comparison similarly to the previous case. However, even in this case, DTC resolution must be finer than the standard deviation of random noise, in order to avoid unwanted regrowth of fractional spurs and quantization noise.

5.3.4 **Fractional-*N* divider with unregulated DTC**

The DTC-based fractional-*N* divider in the digital PLL with a single-bit TDC can be realized as shown in Figure 5.20. The classical digital $\Delta\Sigma$ modulator generates the signal dithering the modulus of the integer-*N* divider between N and $(N+1)$. The large quantization error produced by this operation is canceled out by means of the DTC. Since the modulus of the divider controls the frequency of its output while the DTC controls directly the phase of its output signal, an integration of the $\Delta\Sigma$ quantization error must be performed in the digital domain before applying it to the DTC control $a[k]$.

The required high-resolution DTC could be in principle realized as a phase-interpolator circuit made of cascaded delay stages. Unfortunately, assuming a flash-type approach, a 10-bit DTC would require more than 10^3 phases, thus precluding the use of a delay line as well as the realization of a multiphase oscillator. Since only one phase needs to be generated and compared to the reference signal at a certain clock cycle, the DTC based

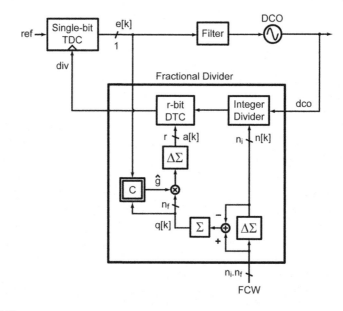

FIGURE 5.20

Digital PLL with unregulated DTC and single-bit TDC.

on the digitally controlled delay stage in Figure 5.9 can be employed. The conversion characteristic of this *unregulated* DTC is schematically shown in Figure 5.21. The output range of the DTC is not exactly equal to T_{dco} but it is instead deliberately designed to be larger than T_{dco}. Assuming the digital gain, $\hat{g}$, is equal to 1, the phase ramp at the output of the divider in Figure 5.20 would show the periodic discontinuities schematically illustrated in Figure 5.22a and resulting in large fractional spurs.

For this reason, the $q[k]$ signal driving the DTC is first multiplied by the proper $\hat{g}$ and quantized by a $\Delta\Sigma$ modulator, so to adapt the DTC output range to T_{dco}. In this way, the range used in the DTC exactly matches the delay induced by toggling the modulus of the integer-N divider from N to $(N+1)$. The resulting phase ramp illustrated in Figure 5.22b is continuous and the quantization error with respect to an ideal continuous phase ramp with constant slope is proportional to the DTC resolution (thus much lower) and it has mainly high-frequency content (filtered out by the loop) since it is generated by a $\Delta\Sigma$ modulator.

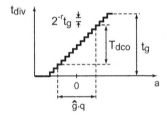

FIGURE 5.21

Conversion characteristic of the unregulated DTC (output delay vs. input code).

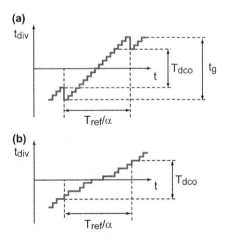

FIGURE 5.22

Phase ramp of the fractional divider in Figure 5.20: (a) before and (b) after the digital regulation of the DTC.

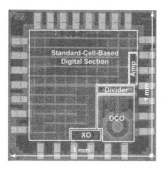

FIGURE 5.23

Die photograph of the digital PLL with single-bit TDC in [52].

The gain, $\hat{g}$, is automatically regulated in the background by means of an adaptive filter. The desired value of the gain, $\hat{g}$, is obtained by means of sign-least-mean-square (sign-LMS) algorithm [51]. The signal, $q[k]$, to be canceled out is correlated with the single-bit output $e[k]$ of the TDC by means of a simple multiplication, and the result is integrated to get $\hat{g}$. In practice, with this method, the DTC regulation is fully performed in the digital domain with much lower power consumption if compared to an analog regulation.

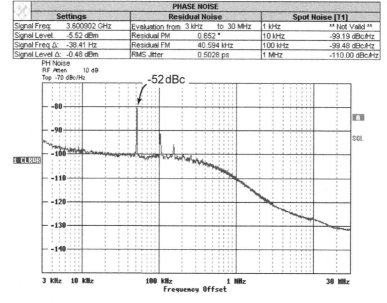

FIGURE 5.24

Measured spectrum of the digital PLL in [52] for a near-integer channel.

The 3.6 GHz digital PLL in [49] implementing such scheme reports a 20 dB reduction in the level of fractional spurs. The worst-case spur obtained for near-integer channels has a residual level of $-42\,$dBc, which is likely caused by residual DTC nonlinearity and spurious coupling between sensitive blocks. The absolute jitter achieved with a loop bandwidth of 300 kHz is 420 fs (or equivalently $-40\,$dBc integrated phase noise), when spurs fall out of the bandwidth, and it increases to 563 fs (or $-38\,$dBc), when near-integer channels are synthesized and fractional spurs fall in-band. Given the low PLL dissipation of only 4.5 mW, obtained thanks to the adoption of the low-power single-bit TDC, that work achieves the best jitter/power trade-off reported today among published analog and digital fractional-N synthesizers.

An improved version of this PLL presented in [52] adds the digital correction of the DTC nonlinearity and the capability of modulating the carrier at an extreme bandwidth of 20 Mb/s. In this case, the test chip in 65 nm CMOS shown in Figure 5.23 demonstrates a maximum level of in-band spurs of about $-52\,$dBc and a worst-case absolute jitter of 503 fs (see spectrum in Figure 5.24), while dissipating 5 mW. In this case, the digital section besides the previously described algorithms includes the generation of the modulation signals and several other correction techniques (described in [52]), which allows reaching an error vector magnitude (EVM) of 36 dB when modulating the carrier with a QPSK at 20 Mb/s.

CONCLUSIONS

Even though the first digital implementations of PLLs date back to the 1960s, only in the last decade has the scaling of CMOS processes allowed them to reach sufficiently low-noise and low-spur performance to be applicable as local oscillators in radio applications. Digital PLLs offer several advantages over their analog counterparts, but their design must face some challenges. In this chapter, we discussed the effects of TDC finite resolution and nonlinearity, which affects the spur performance, especially in wideband PLLs. The main algorithms for mitigating TDC nonlinearity such as those based on element randomization and large-scale dithering have been reviewed. An alternative approach consists of revising the digital PLL architecture and introducing fractional-N dividers based on DTCs as a means to relax the design of the time-to-digital converter. This concept has been illustrated in a practical case and then it has been extended to the limit of a single-bit time-to-digital converter, which provides the best PLL noise–power trade-off with good spur performance.

References

[1] R. B. Staszewski et al., All-digital TX frequency synthesizer and discrete-time receiver for bluetooth radio in 130-nm CMOS, *IEEE Journal of Solid-State Circuits*, vol. 39, no. 12, pp. 2278–2291, December 2004.

[2] C.-M. Hsu, M. Z. Straayer, and M. H. Perrott, A low-noise wide-BW 3.6-GHz digital $\Delta\Sigma$ fractional-N frequency synthesizer with a noise-shaping time-to-digital converter

and quantization noise cancellation, *IEEE Journal of Solid-State Circuits*, vol. 43, no. 12, pp. 2776–2786, December 2008.

[3] E. Temporiti et al., A 700-kHz bandwidth ΣΔ fractional synthesizer with spurs compensation and linearization techniques for WCDMA applications, *IEEE Journal of Solid-State Circuits*, vol. 39, no. 9, pp. 1446–1454, September 2004.

[4] M. Gupta and B.-S. Song, A 1.8 GHz spur-cancelled fractional-N frequency synthesizer with LMS-based DAC gain calibration, *IEEE International Solid-State Circuits Conference. Digest of Technical Papers*, pp. 1922–1923, February 2006.

[5] G. W. Roberts and M. Ali-Bakhshian, A brief introduction to time-to-digital and digital-to-time converters, *IEEE Transactions on Circuits and Systems–II: Express Briefs*, vol. 57, no. 3, pp. 153–157, March 2010.

[6] J.-P. Jansson, A. Mantyniemi, and J. Kostamovaara, A CMOS time-to-digital converter with better than 10 ps single-shot-precision, *IEEE Journal of Solid-State Circuits*, vol. 41, no. 6, pp. 1286–1296, June 2006.

[7] M. Lee, M. Heidari, and A. A. Abidi, A low-noise wideband digital phase-locked loop based on a coarse-fine time-to-digital converter with subpicosecond resolution, *IEEE Journal of Solid-State Circuits*, vol. 44, no. 10, pp. 2808–2816, April 2009.

[8] M. Z. Straayer and M. H. Perrott, A multi-path gated ring oscillator TDC with first-order noise shaping, *IEEE Journal of Solid-State Circuits*, vol. 44, no. 4, pp. 1089–1098, April 2009.

[9] L. Vercesi, A. Liscidini, and R. Castello, Two dimensions Vernier time-to-digital converter, *IEEE Journal of Solid-State Circuits*, vol. 45, no. 8, pp. 1504–1512, August 2010.

[10] J. Borremans, K. Vengattaramane, V. Giannini, and J. Craninckx, A 86 MHz–12 GHz digital-intensive PLL for software-defined radios, using a 6 fJ/Step TDC in 40 nm digital CMOS, *IEEE Journal of Solid-State Circuits*, vol. 45, no. 10, pp. 2116–2129, October 2010.

[11] T. Tokairin et al., A 2.1-to-2.8-GHz low-phase-noise all-digital frequency synthesizer with a time-windowed time-to-digital converter, *IEEE Journal of Solid-State Circuits*, vol. 45, no. 12, pp. 2582–2590, December 2010.

[12] D. Lee, J. Han, G. Han, and S. M. Park, A 1 GHz ADPLL with a 1.25 ps minimum-resolution sub-exponent TDC in 0.18m CMOS, *IEEE Journal of Solid-State Circuits*, vol. 45, no. 12, pp. 2874–2882, December 2010.

[13] A. Elshazly, S. Rao, B. Young, and P. K. Hanumolu, A 13b 315fsrms 2 mW 500 MS/s 1MHz bandwidth highly digital time-to-digital converter using switched ring oscillators, *IEEE International Solid-State Circuits Conference. Digest of Technical Papers*, pp. 464–466, February 2012.

[14] P. Lu, A. Liscidini, and P. Andreani, A 3.6 mW, 90 nm CMOS gated-Vernier time-to-digital converter with an equivalent resolution of 3.2 ps, *IEEE Journal of Solid-State Circuits*, vol. 47, no. 7, pp. 1626–1635, July 2012.

[15] F. M. Gardner, Charge-pump phase-locked loops, *IEEE Transactions on Communications*, vol. 28, no. 11, pp. 1849–1858, November 1980.

[16] E. Temporiti et al., A 3 GHz fractional all-digital PLL with a 1.8 MHz bandwidth implementing spur reduction techniques, *IEEE Journal of Solid-State Circuits*, vol. 44, no. 3, pp. 824–834, March 2009.

[17] C. Weltin-Wu et al., Insights into wideband fractional ADPLLs: modeling and calibration of nonlinearity induced fractional spurs, *IEEE Transactions on Circuits and Systems—I: Regular Papers*, vol. 57, no. 9, pp. 2259–2268, September 2010.

[18] M. Zanuso, S. Levantino, C. Samori, and A. L. Lacaita, A wideband 3.6 GHz digital $\Delta\Sigma$ fractional-N PLL with phase interpolation divider and digital spur cancellation, *IEEE Journal of Solid-State Circuits*, vol. 46, no. 3, pp. 627–638, March 2011.

[19] Ja-YoiLee, Mi-JeongPark, Byung-HunMin, SeongdoKim, Mun-YangPark, and Hyun-KyuYu, A 4-GHz All Digital PLL with low-power TDC and phase-error compensation, *IEEE Transactions on Circuits and Systems—I: Regular Papers*, vol. 59, no. 8, pp. 1706–1719, August 2012.

[20] L. Xu, K. Stadius, and J. Ryynanen, An all-digital PLL frequency synthesizer with an improved phase digitization approach and an optimized frequency calibration technique, *IEEE Transactions on Circuits and Systems—I: Regular Papers*, vol. 59, no. 11, pp. 2481–2494, November 2012.

[21] F. Gardner, Frequency granularity in digital phaselock loops, *IEEE Transactions on Communications*, vol. 44, no. 4, pp. 749–758, June 1996.

[22] P. Madoglio, M. Zanuso, S. Levantino, C. Samori, and A. L. Lacaita, Quantization effects in all-digital phase-locked loops, *IEEE Transactions on Circuits and Systems—II: Express Briefs*, vol. 54, no. 12, pp. 1120–1124, December 2007.

[23] K. Takinami et al., A distributed oscillator based all-digital PLL with a 32-phase embedded phase-to-digital converter, *IEEE Journal of Solid-State Circuits*, vol. 46, no. 11, pp. 2650–2660, November 2011.

[24] L. Vercesi et al., A dither-less all digital PLL for cellular transmitters, *IEEE Journal of Solid-State Circuits*, vol. 47, no. 8, pp. 1908–1920, August 2012.

[25] E. Temporiti et al., A 3.5 GHz wideband ADPLL with fractional spur suppression through TDC dithering and feedforward compensation, *IEEE Journal of Solid-State Circuits*, vol. 45, no. 12, pp. 2723–2736, December 2010.

[26] T. Maeda and T. Tokairin, Analytical expression of quantization noise in time-to-digital converter based on the Fourier series analysis, *IEEE Transactions on Circuits and Systems—I: Regular Papers*, vol. 57, no. 7, pp. 1538–1548, July 2010.

[27] M. Zanuso et al., Time-to-digital converter with 3-ps resolution and digital linearization algorithm, *Proceedings of the European Solid-State Circuits Conference ESSCRIC 2010*, pp. 262–265, September 2010.

[28] V. Kratyuk et al., A digital PLL with a stochastic time-to-digital converter, *IEEE Transactions on Circuits and Systems—I Regular Papers*, vol. 56, no. 8, pp. 1612–1621, August 2009.

[29] R. Staszewski et al., Elimination of spurious noise due to time-to-digital converter, *IEEE Dallas Circuits and Systems Workshop*, October 2009.

[30] R. M. Gray and T. G. StockhamJr , Dithered quantizers, *IEEE Transactions on Information Theory*, vol. 39, pp. 805–812, May 1993.

[31] S. P. Lipshitz, R. A. Wannamaker, and J. Vanderkooy, Quantization and dither: a theoretical survey, *Journal of the Audio Engineering Society*, vol. 40, pp. 355–375, May 1992.

[32] B. Murmann, A/D converter trends: power dissipation, scaling and digitally assisted architectures, in *Proceedings of the IEEE 2008 Custom Integrated Circuits Conference (CICC)*, pp.105–112, September 2008.

[33] C. Cottini, E. Gatti, and V. Svelto, A new method for analog to digital conversion, *Nuclear Instruments and Methods*, vol. 24, pp. 241, August 1963.

[34] P.-Y. Wang, J.-H.C. Zhan, H.-H. Chang, and H.-M.S. Chang, A digital intensive fractional-N PLL and all-digital self-calibration schemes, *IEEE Journal of Solid-State Circuits*, vol. 44, no. 8, pp. 2182–2192, August 2009.

[35] C.-H. Park, O. Kim, and B. Kim, A 1.8-GHz self-calibrated phase-locked loop with precise I/Q matching, *IEEE Journal of Solid-State Circuits*, vol. 36, no. 5, pp. 777–783, May 2001.

[36] S. Pamarti and S. Delshadpour, A spur elimination technique for phase interpolation-based fractional-N PLLs, *IEEE Transactions on Circuits and Systems—I: Regular Papers*, vol. 55, no. 6, pp. 1639–1647, July 2003.

[37] P.-E. Su and S. Pamarti, A 2-MHz bandwidth $\Delta-\Sigma$ fractional-N synthesizer based on a fractional frequency divider with digital spur suppression, *Proceedings of IEEE Radio Frequency Integrated Circuits Symposium RFIC 2010*, pp. 413–416, May 2010.

[38] M. Zanuso et al., Time-to-digital converter for frequency synthesis based on a digital bang-bang DLL, *IEEE Transactions on Circuits and Systems I—Regular Papers*, vol. 57, no. 3, pp. 548–555, March 2010.

[39] M. S-W. Che, D. Su, and S. Mehta, A calibration-free 800 MHz fractional-N digital PLL with embedded TDC, *IEEE Journal of Solid-State Circuits*, vol. 45, no. 12, pp. 2819–2827, December 2010.

[40] D. Miyashita et al., A −104 dBc/Hz in-band phase noise 3 GHz all digital PLL with phase interpolation based hierarchical time to digital convertor, *Digest of Technical Papers of VLSI Symposium on Circuits*, pp. 112–113, June 2011.

[41] C. Samori, M. Zanuso, S. Levantino, and A. L. Lacaita, Multipath adaptive cancellation of divider non-linearity in fractional-N PLLs, *Proceedings of the IEEE International Symposium on Circuits and Systems (ISCAS) 2011*, pp. 418–421, May 2011.

[42] A. Rylyakov et al., Bang-bang digital PLLs at 11 and 20 GHz with sub-200fs integrated jitter for high speed serial communication applications, *IEEE International Solid-State Circuits Conference. Digest of Technical Papers*, No. 12, pp. 94–96, February 2009.

[43] M. Ferriss and M. Flynn, A 14 mW fractional-N PLL modulator with a digital phase detector and frequency switching scheme, *IEEE Journal of Solid-State Circuits*, vol. 43, no. 11, pp. 2464–2471, November 2008.

[44] D.-W. Jee, Y.-H. Seo, H.-J. Park, and J.-Y. Sim, A 2 GHz fractional-N digital PLL with 1b noise shaping $\Delta\Sigma$ TDC, *IEEE Journal of Solid-State Circuits*, vol. 47, no. 4, pp. 875–883, April 2012.

[45] M. D. McDonnell, Is electrical noise useful? *Proceedings of the IEEE*, vol. 99, no. 2, pp. 242–246, February 2011.

[46] N. Da Dalt, Markov chains-based derivation of the phase detector gain in bang-bang PLL, *IEEE Transactions on Circuits and Systems—II: Exp Briefs*, vol. 53, no. 11, pp. 1195–1199, November 2006.

[47] N. Da Dalt, Linearized analysis of a digital bang-bang PLL and its validity limits applied to jitter transfer and jitter generation, *IEEE Transactions on Circuits and Systems—I: Regular Papers*, vol. 55, no. 11, pp. 3663–3675, November 2008.

[48] M. Zanuso et al., Noise analysis and minimization in bang-bang digital PLL, *IEEE Transactions on Circuits and Systems—I: Regular Papers*, vol. 56, no. 11, pp. 835–839, November 2009.

[49] D. Tasca et al., A 2.9-to-4.0 GHz fractional-N digital PLL with bang-bang phase detector and 560fsrms integrated jitter at 4.5 mW power, *IEEE Journal of Solid-State Circuits*, vol. 46, no. 12, pp. 2745–2758, December 2011.

[50] N. Pavlovic and J. Bergevoert, A 5.3 GHz digital-to-time-converter-based fractional-N all-digital PLL, *IEEE International Solid-State Circuits Conference. Digest of Technical Papers*, pp. 54–55, February 2011.

[51] A. H. Sayed, *Adaptive Filters*. Wiley-IEEE Press, 2008.

[52] G. Marzin, S. Levantino, C. Samori, and A. L. Lacaita, A 20 Mb/s phase modulator based on a 3.6 GHz digital PLL with −36 dB EVM at 5 mW power, *IEEE International Solid-State Circuits Conference. Digest of Technical Papers*, pp. 342–343, February 2012.

Mixers and Modulators in Wireless Systems

6

William Redman-White

Electronics and Computer Science, University of Southampton,
Southampton SO17 1BJ, UK

INTRODUCTION

Mixers (or modulators) are some of the most important functions in a radio transceiver. Almost all radio architectures use mixers to convert the carrier frequency of a wanted signal from one range to another for various reasons. The requirement for a mixer was first identified with the development of the superheterodyne architecture, wherein it was shown to be advantageous to translate the incoming RF to a lower frequency in order to make the required channel filtering practical and the main amplification more efficient. These motivations still hold good today.

The mixer is also one of the most critical functions in most receivers inasmuch as it is sufficiently close to the input to have an impact on noise figure, and at the same time, is usually required to have very high linearity in order to process signals in a moderately wide bandwidth prior to channel filtering.

While the mathematical function performed is broadly the same in most cases, the precise implementation used will depend on a range of factors, including the type of radio architecture used, the position of a specific mixer in the signal flow, the performance needed, the semiconductor technology in use, etc.

6.1 Basic principles

At heart, all mixers perform the same basic mathematical function in order to achieve the translation of the carrier frequency of a signal while retaining the modulation information carried thereon; and this is multiplication. When a wanted signal, be it in the transmit or receiver path, and represented, for simplicity, by a single sinusoid $A_{IN}\sin(\omega_{IN} t)$, is multiplied by the local oscillator (LO) signal sinusoid $A_{LO}\sin(\omega_{LO} t)$, the resulting output contains terms with the sum and difference frequencies (Figure 6.1).

Historically, it was common to use a nonlinear mixer, in which the wanted signal and the LO are fed into the input port of a stage having some nonlinear response. Provided that the stage nonlinearity contains a significant quadratic term, multiplication terms are generated in the second-order products. This is seldom used for

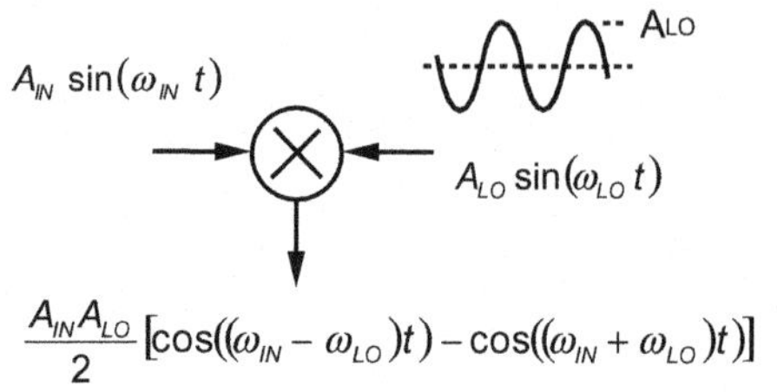

FIGURE 6.1

Ideal multiplier as a mixer.

communications due to the near-certainty of severe intermodulation, but the mechanism still poses a risk of unwanted mixing in the front-end of zero-IF receivers.

In the case of an ideal multiplication operation, the sum and difference frequencies are the only terms present in the output, but this is seldom actually the case. The nature of the other products in a practical circuit will depend on the specific implementation used. An ideal multiplier circuit would appear to be a good choice, but this has implications for system noise.

6.2 Switching mixers

The mixer operation employed in virtually all modern radio systems is that of switching, whereby the polarity of the wanted signal is reversed in accordance with the instantaneous sign of the LO signal. This signal reversing operation is equivalent to multiplying the wanted signal by a square wave (Figure 6.2) whose frequency is ω_{LO}. This means that the LO will naturally have Fourier components at (ideally) odd multiples of ω_{LO}. As a consequence, the incoming signal is also multiplied by these Fourier components, also resulting in unwanted sidebands around these terms with reducing amplitude.

In most receiver systems the responses to harmonics of the LO are not a problem, as there is almost always a low-pass or band-pass filter following the mixer that will easily reject the unwanted terms.

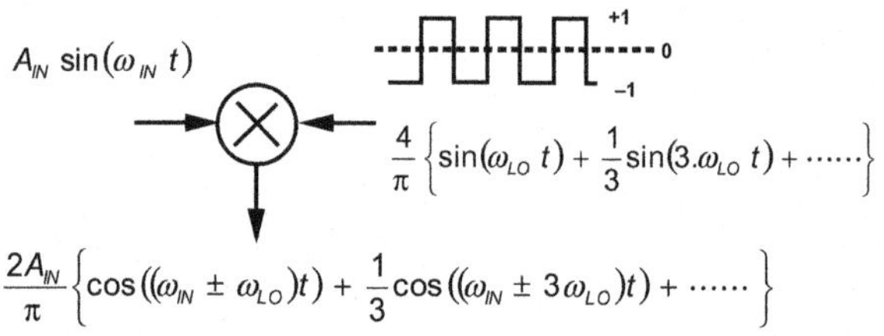

FIGURE 6.2

Switching circuit as a mixer.

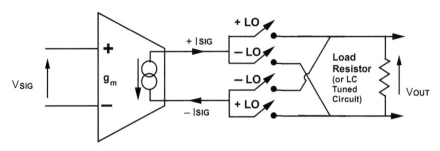

FIGURE 6.3

General behavioral model of a switching mixer.

The operation is most easily performed by taking the incoming signal as a current and then reversing the polarity with some switching elements (Figure 6.3).

In addition to much-reduced circuit complexity, a switching mixer has major advantages for system noise figure. The circuit level issues will be discussed later. At a system level, it will be seen that an ideal switching process also removes any amplitude information from the local oscillator (LO) signal, thereby reducing the noise contribution of the LO to only the phase term.

6.3 Specifications

The most commonly quoted performance metrics are the conversion gain, noise figure, IIP3 and IIP2 [1].

6.3.1 Gain

The conversion gain is classically defined in a matched system as

$$G_C = \frac{\text{Power in Wanted Output Freq. to Load}}{\text{Power in Signal from Source}}.$$

In an IC context, the gain is best considered in voltage terms, and hence is defined as the ratio of the rms voltage in the input signal to the rms output of the wanted sideband. As can be seen, operation following the ideal switching behavioral model leads to a gain of $2/\pi$ for each sideband.

6.3.2 Linearity

The third-order linearity behavior of a mixer can be viewed broadly in the same way as for an amplifier, except that the spectrum containing the wanted signal and intermodulation spectra will be translated to appear around the new carrier frequency at the output (Figure 6.4). This complicates the simulation procedures and requires some care in the setting of the simulator control parameters for periodic steady-state analyses if excessively long computations are to be avoided.

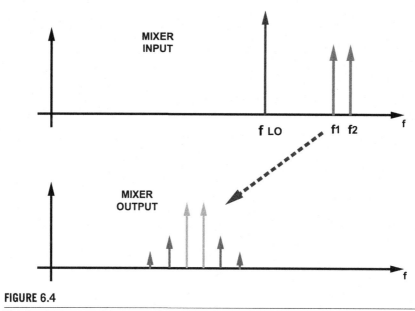

FIGURE 6.4

Third-order intermodulation products in a mixing process.

Published papers sometimes still quote IIP3 numbers in dBm for a mixer cell alone, with the implied reference to a $50\,\Omega$ source. This is less useful in the context of an integrated design, but the measure has some value for the overall system IIP3, where the mixer may be the dominant nonlinearity due to the preceding gain of the LNA.

The second-order intermodulation figure has become increasingly important with the widespread adoption of zero and low IF architectures. This is because with any second-order nonlinearity in the mixer input stage, two unwanted signals with closely spaced frequencies can give rise to a difference term that falls within the channel filter bandwidth at the mixer's output (Figure 6.5) [2,3]. Very high-input referred second-order linearity is required for a successful ZIF design.

Such second-order nonlinearity originating in a preceding amplifier stage is less likely to be a problem due to the high-pass nature of most interstage coupling arrangements.

6.3.3 Leakage and feedthrough

In almost all practical mixer implementations there is some degree of leakage from one port to another (Figure 6.6). In some cases this is of no consequence, but in some applications it is critical.

For ZIF receivers, leakage from the LO port to the input is the most well-known risk. The mechanism for this is mainly related to the capacitive coupling across some of the internal devices, even in an otherwise ideal circuit. In-band radiation may occur in a low-specification receiver with no LNA to isolate the antenna, but a more troublesome effect is self-mixing of the LO entering the signal port that gives rise to DC offsets in

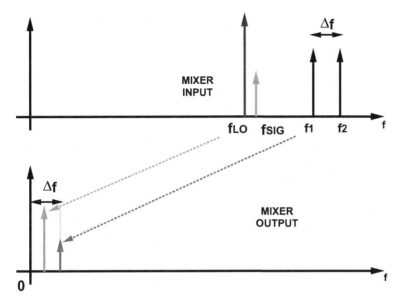

FIGURE 6.5

Second-order intermodulation mixing products in a ZIF or LIF receiver.

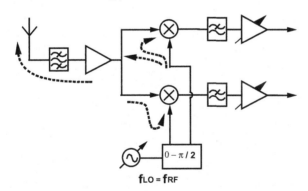

FIGURE 6.6

Mixer leakage and feed-through paths in a receiver front end.

the receive chain [2]. Leakage from the signal port to the LO port can also be a problem, as this can lead to second-order intermodulation terms appearing at the output [4].

Leakage from the signal and LO to the output port can be due to imbalances and other imperfections [4,5]. If the LO signal does not have ideal symmetry, there will be an additional DC term in the LO spectrum, multiplying the input signal:

$$V_{\text{LONonIdeal}}(t) = \left(d - \frac{1}{2}\right) + \frac{4}{\pi}\left\{\begin{array}{l}\sin(\pi d)\sin(\omega_{\text{LO}}t) + \sin(2\pi d)\sin(2\omega_{\text{LO}}t) \\ + \sin(3\pi d)\sin(3\omega_{\text{LO}}t) + \cdots\end{array}\right\},$$

where *d* is the duty cycle, ideally 0.5. Any imbalance leads to a leakage signal component at the output (as well as weak sidebands around the odd LO harmonics). This is not usually very serious in itself, due to subsequent filtering, but does lead to a path to the output for any IM2 products created in the input voltage-to-current conversion function, and hence LO symmetry should be observed. Note that any DC drift in the LO buffer due to LO asymmetry and AC coupling has been ignored here on the assumption that the LO waveform is still large enough to achieve full switching at each excursion, and transition times are short.

Leakage from the LO port to the output is mostly driven by offsets in the DC bias passing through the mixer. This is particularly troublesome in transmitters as the unwanted carrier component is usually too close to the signals to be filtered.

6.3.4 **Noise**

The concept of noise figure is generally of concern in receiver applications, as the mixer often follows only a very limited gain, and hence its noise figure can still be significant in the overall system noise figure. The issue is complicated by the need to consider the receiver architecture used [1]. In a traditional superhet the noise apparent at the output has three distinct sources. Firstly, there is the noise due to the internal circuitry of the mixer, usually dominated by the voltage-to-current conversion function. Secondly, there is the noise present at the input port within the signal bandwidth. Finally, there is also noise in the image band that mixes with the LO to translate noise to the output IF band. This is the case even if there is an image rejection filter preventing any signal entering this part of the spectrum. If the frequency

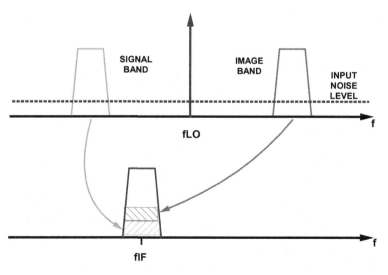

FIGURE 6.7

Origins of single sideband-noise figure.

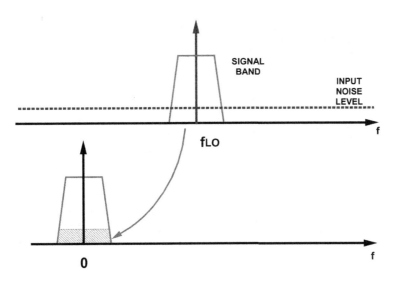

FIGURE 6.8

Origins of double-sideband noise figure.

response of the mixer is such that the gain for the image band is similar to that in the wanted signal band, then the noise apparent at the output will be twice that for the signal band alone, leading to a noise figure of at least 3 dB, even if the internal circuitry is noiseless (Figure 6.7). This is referred to as the single-sideband (SSB) noise figure.

The alternative use case scenario is in a zero-IF (ZIF) receiver. In this situation, there is only one segment of the frequency spectrum that has noise converted to the output, and hence the noise figure will be determined by the noise at the input with respect to the noise in the output IF channel, with the main contribution being the internal circuitry (Figure 6.8). This is referred to as the double sideband (DSB) noise figure.

Some care needs to be taken when assessing the overall system noise figure in a ZIF receiver, as the two signal paths will be combined in the baseband signal processing, with different results for the summation of the different signal and noise contributions [1].

6.4 Single and double balanced mixers

The ideal switching operation described above gives rise only to output components that are at the sum and difference of the LO and signal frequencies (and, in most cases, the sum and difference of the signal and the LO harmonics). Neither the input nor the LO signals are apparent in the output spectrum in this ideal case. Such a mixer is referred to as a double balanced mixer (a description derived from the form of the classical transformer-driven diode ring mixer structure).

It is also possible to have a mixer where one of the input signals is present in the output, usually the LO component, and this is referred to as a single balanced mixer. This is usually due to the presence of a DC bias current flowing through the switching core, giving rise to a term at the LO frequency.

$$(A_{DC} + A_S \sin(\omega_S t)) \times \sin(\omega_{LO} t) = \frac{A_S}{2} \{\cos((\omega_S - \omega_{LO})t) - \cos((\omega_S + \omega_{LO})t)\}$$
$$+ A_{DC} \sin(\omega_{LO} t) \quad \text{(ignoring the LO harmonics)}.$$

While seemingly not as desirable as the double balanced form, the single balanced mixer can still be of value. In a receiver architecture, there is almost always a low-pass or band-pass filter following the mixer, and hence an output component at the LO frequency is easily removed. In a simple architecture, there can be a saving in the power consumption of the mixer function compared with the double balanced form, both in terms of the signal path and the LO drive requirements.

However, in a transmitter, the single balanced form is not usually suitable as an output component due to the up-conversion of the LO being difficult to remove. In a direct up-conversion scheme this will lead to an LO component in the center of the signal band, and even in a superhet transmitter, any RF filtering in the transmit path is, of necessity, relatively simple in form. This is dictated by the difficulties in handling relatively strong signals with active circuitry, and it is also apparent that filtering out unwanted frequency components from a high-level signal represents a significant waste of battery power. As a consequence, it is usual to use double balanced mixers in quadrature with a sideband suppression scheme.

6.4.1 Practical implementations

Most integrated circuit implementations fall into two main categories, i.e. *Passive* and *Active* mixers. Both implement the behavioral model of a transconductor followed by reversing switches in the output current path, as per Figure 6.3. The main distinction is that in an active mixer there is a DC bias flowing through the switches, while in the passive form there is none. As a rough generalization, the passive mixer form is best suited to CMOS/BiCMOS receiver applications, while the active type is suited to BiCMOS/bipolar technology in the receive path, and CMOS/BiCMOS/bipolar in the transmit path.

6.5 Passive mixers

A typical arrangement is shown in Figure 6.9. The LNA is used as a transconductor and the output is taken directly as a current. The LNA may be any classical form (common source with cascode, or inverter topology), but a high output impedance is desired. A small coupling capacitor between the LNA and the switching core isolates the MOS switches from the LNA bias point, and serves to block low-frequency

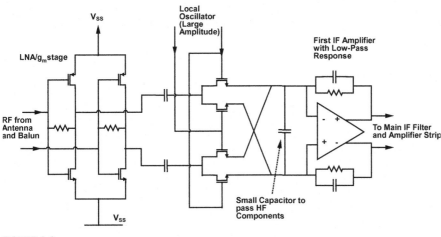

FIGURE 6.9

Typical current-mode passive mixer arrangement.

IM2 products from the LNA that might pass through the mixer if any imbalance exists [6]. The MOS reversing switches are typically small NMOS transistors, and these are connected directly to the virtual ground of a transimpedance amplifier, so that the common mode level of the switches is set by the amplifier's CMFB loop. The LO signals are large (normally rail-to-rail amplitude) to achieve a low *ON* resistance in the switches. With the low impedance at the amplifier input, and the high impedance at the LNA output, the switch resistances are largely constant and independent of the RF signal, greatly contributing to high linearity. A voltage output is taken from the transimpedance amplifier, which typically includes a first real pole in the baseband filtering. A small capacitor at the input of the amplifier is usually used to allow currents to circulate at high frequencies where the amplifier's feedback has rolled off.

The transimpedance amplifier must have very low noise as well as high enough gain to set the virtual ground effectively; as its bandwidth is necessarily limited, this type of mixer is generally usable only as a down-conversion mixer in a receiver.

6.6 Passive mixer noise sources

A major advantage of this type of mixer is that the absence of any DC bias through the MOS switches leads to a low transconductance for the switching core devices viewed from the LO ports, and their noise contributions, and of particular concern, the 1/f noise, can be kept very small (although not actually zero) [7,8], making the architecture very attractive for CMOS ZIF receivers [3].

In practice, the main problem is the input referred noise of the transimpedance amplifier, with a time-dependent gain factor due to the effective input network.

The precise nature of this gain depends on the LO waveforms, the switch threshold voltages, and the input common mode level of the transimpedance amplifier (Figure 6.10) [7].

If the LO waveforms are such that the core MOS switches are all turned on at the same time, then it follows that there is a path completed for the amplifier noise to flow at the input (Figure 6.11).

Hence the amplifier input noise is magnified by a factor of the order of

$$V_{\text{out}} \approx V_{\text{in}} \left(1 + \frac{2R_{\text{FB}}}{R_{\text{SW}}} \right).$$

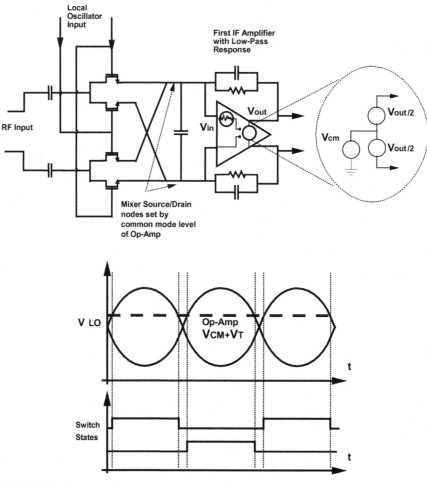

FIGURE 6.10

Noise sources and LO switching instants in relation to amplifier common mode level.

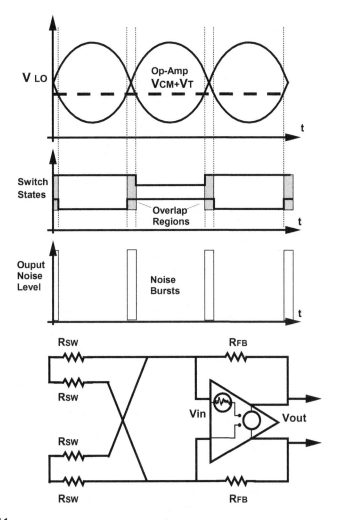

FIGURE 6.11

Amplifier noise magnification due to overlap of LO switching instants.

Given that the switch resistances are deliberately made very low to achieve good linearity, and the feedback resistances will be relatively high for good conversion gain, this noise amplification factor can be very large, leading to a burst of noise at the crossover of the LO signals.

As a consequence, the LO signals must be defined with care with respect to the amplifier input common mode level and the MOS thresholds to ensure that this loop cannot form; effectively a break-before-make sequence is required. The amplifier common mode level is usually constrained by the need to design for a very low input referred noise level, particularly at low frequencies, and hence the DC level for the LO signals is derived from this reference combined with the threshold voltage of the

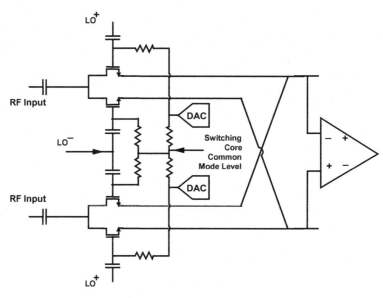

FIGURE 6.12

Calibration of switching instants to minimize feedthrough by adjusting individual LO DC pedestals.

MOS switches to ensure no overlap of the *ON* phases. A convenient way to achieve this is to generate the LO signals in rail-to-rail CMOS circuits and then capacitively couple to the switching core gates with the common mode bias supplied via high value resistances. This can be further enhanced by the use of a trimming DAC system (Figure 6.12) to allow the effective duty cycles of the opposing phases to be balanced, reducing feedthrough from input to output ports, and hence reduce IM2 products arising in the input stages [9].

In addition to this ohmic path at the input, there is also a path for noise current to flow via the switched capacitances at the input node due to the coupling capacitor and various device drain-substrate capacitances. These should also be minimized. When using this type of mixer for separate I-Q branches it is also necessary to connect each branch to the LNA g_m output with separate coupling capacitors to avoid the transimpedance stage noise in the I path from circulating to the Q path virtual ground and vice versa [10].

6.7 25% Duty cycle passive mixer

When used in a zero- or low-IF receiver with I and Q baseband paths, the performance of a passive mixer can be further improved by the use of 25% duty cycle LO signals (Figure 6.13) [11,12]. In addition to eliminating the path for amplifier noise current between the two branches [10], this arrangement has the advantage that at any given

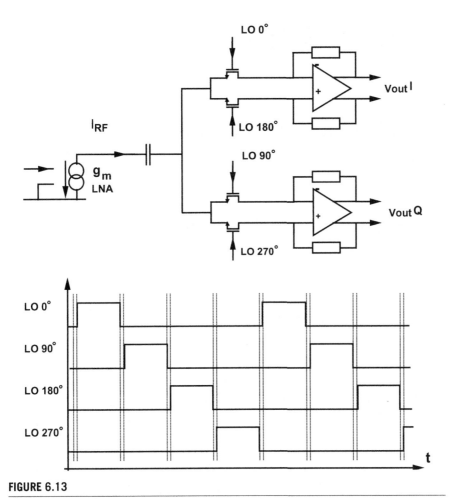

FIGURE 6.13

Passive mixer for complex receive signal paths using non-overlapping 25% duty cycle LO waveforms.

time the output of the LNA is only feeding a single mixer, and is not shared [13]. If we define the LNA transconductance as g_m, and the transimpedance stage as R_m, then with a 50% duty cycle we can derive the approximate conversion gain for the I path, noting that the fundamental Fourier coefficient of a square wave is $4/\pi$ with a further factor of 1/2 from the multiplication as before; but we must note that as the Q path switches will be on at the same time, then only half the LNA current is available:

$$G_{C(50\%)} = g_m \times \frac{1}{2} \times \frac{4}{\pi} \times \frac{1}{2} \times R_m = \frac{g_m R_m}{\pi}$$

where 25% duty cycle LO signals are used, we note that the corresponding fundamental Fourier coefficient is $\frac{2\sqrt{2}}{\pi}$, again with a factor of 1/2 from the multiplication. However, in this case all of the LNA current is available, so the conversion gain is:

$$G_{C(25\%)} = g_m \times \frac{2\sqrt{2}}{\pi} \times \frac{1}{2} \times R_m = \frac{\sqrt{2}g_m R_m}{\pi}.$$

With the same circuit elements, this improves the signal-to-noise ratio by nominally 3 dB.

There is a small penalty insofar as the LO generation requires a more complicated (and hence more power hungry) logic block to generate the waveforms with sufficient accuracy, but the noise advantages are making the 25% passive mixer increasingly popular in low- and zero-IF CMOS receivers.

6.8 Active mixers

The most common configuration in IC form has traditionally been the "active" current steering (often referred to as a "Gilbert" mixer) [14]. This also follows the basic configuration of a linear transconductor, followed by steering switches driven differentially by the LO. At the output some load impedance converts the signal back into a voltage.

The conversion gain is easily derived from the transconductance and load values, scaled by $2/\pi$.

Figure 6.14 shows the basic single balanced current steering mixer in bipolar form. As can be seen, there will always be a constant bias current flowing along with the signal component, and this is an important factor in the analysis of the noise sources. The signal and bias currents are steered between the two load impedances by means of a differential pair arrangement.

The double balanced form (Figure 6.15) is by far the most often used, both in bipolar and MOS forms [15]. With a bias current in both halves, the summed differential contribution is cancelled out and there is, ideally, no output at the LO frequency.

Depending on the form of the transconductor and the output load impedances, this topology may be used for transmit or receive functions, and can be realized in bipolar or MOS forms.

For a receiver application, the demands of high linearity and wide bandwidth dictate that the transconductor be realized with a simple degenerated common emitter/source arrangement [14], and design follows normal amplifier procedures. If noise figure is a limitation, then inductive degeneration may be employed. In low voltage designs it is often necessary to use a degenerated transconductor with two current sources as shown. This adds to the differential output noise as the tail current sources are uncorrelated, but can improve the available headroom and reduce compression at the output. Note that a mismatch between these two current sources is equivalent to a valid input signal at zero frequency, and hence creates an output at the LO frequency; digital calibration techniques may be used to reduce the offset [16]. In the receive case, the value of the load resistance is chosen to match the noise requirements of the following filter stage, and there is often a passive real pole formed with a single capacitor at this point.

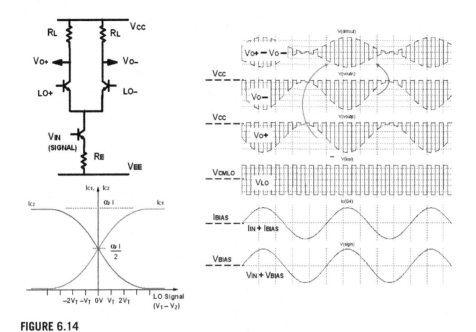

FIGURE 6.14

Single balanced active current steering mixer operation.

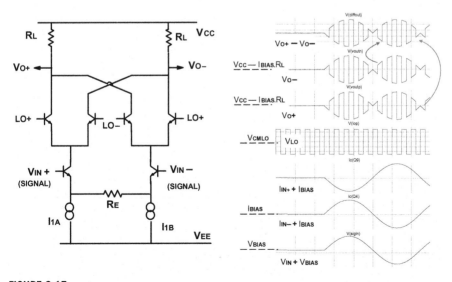

FIGURE 6.15

Double balanced active current steering mixer operation.

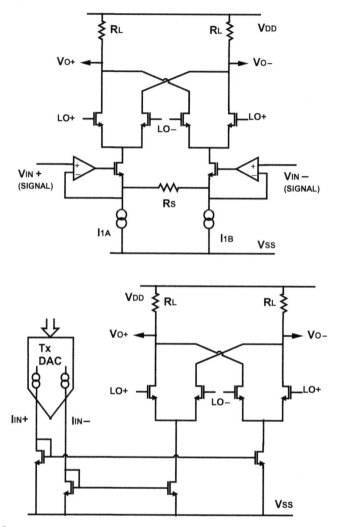

FIGURE 6.16

Active MOS mixers configurations for transmit path, with high linearity V-to-I and direct current DAC input.

This topology is less attractive in a CMOS receiver design. The most serious problem arises from the DC bias present in the switching core, which leads to a significant 1/f noise source. As will be discussed, this noise is reduced by the circuit's operation but is not eliminated, and it is usually unacceptable in a ZIF receiver. In a low-voltage design it is also less easy to use sufficient resistive degeneration to improve the linearity in a simple MOS transconductor [17].

For transmit path applications this topology works well in both MOS and bipolar technology. To achieve high linearity with a large-amplitude baseband input, it is common to use a transconductor with a feedback loop, or to mirror the output from a current DAC directly (Figure 6.16) [18].

The load impedance for a transmit mixer may be resistive or a tuned inductance for higher conversion gain and some filtering of unwanted LO harmonic terms.

6.9 Active mixer noise

A critical issue in the operation of the mixer is the LO drive to the current steering switches, and it is linked to the noise sources in the mixer [19]. The noise in the transconductor clearly adds directly to the signal path and can be treated as in a typical amplifier, but the noise contributed by the differential pair steering switches must be treated differently, as it depends on the region of operation.

When the differential LO drive is sufficient to ensure all the current passes through one side of the pair, Q1, the input referred noise of this device can be treated as though it were in a cascode amplifier, with a degeneration impedance due to the

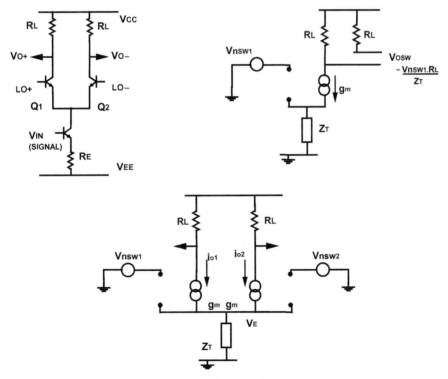

FIGURE 6.17

Noise behavior of the switching core devices in an active mixer when fully switched and during the changeover.

output of the transconductor, Z_{Ogm}. Since this latter term is normally very large, the gain from the input referred noise of Q1 to the output is very low, and usually negligible (Figure 6.17). Some degradation in the attenuation of the switch noise can occur if there is significant capacitance at the output of the transconductor, as this will form a bypass for the noise at high frequencies. This term may be reduced by resonating the parasitic capacitance with an inductor, but at the cost of some area.

When the differential steering transistors are in the middle of the changeover, Q1 and Q2 behave as a conventional differential pair and the device noise contributions of Q1 and Q2 appear at the output with a gain of:

$$\left| \frac{i_{O1}}{V_{nsw1}} \right| = \left| \frac{i_{O2}}{V_{nsw1}} \right| = \frac{g_m}{2}$$

and similarly for Q2. Hence a short burst of noise appears at the mixer output during the changeover (Figure 6.18). This is thermal and shot noise for a bipolar switching core, but will include significant 1/f noise for a MOS implementation in a ZIF receiver [20].

Since this noise is unavoidable during the transition, the obvious strategy is to ensure that the time for the changeover is reduced to a very short value, so that the time averaged noise contribution is minimized.

These signal and noise conditions lead directly to the requirements for the LO drive signal. The goal is to achieve hard switching of the core with the minimum

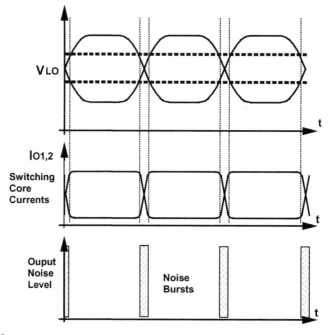

FIGURE 6.18

Noise bursts due to the switching core devices in an active mixer during the changeover.

changeover time, while avoiding other circuit problems. For a bipolar switching core, simple analysis shows that a complete changeover is achieved with a differential LO signal around 200 mV p-p. Since producing an ideal square wave LO at typical radio frequencies is not realistic, an LO amplitude of around 350 mV is a more typical figure. A still larger LO signal is attractive to achieve a faster transition through the changeover region, but the maximum value is limited by the onset of saturation in the switching core transistors as output signal amplitude increases. In a MOS implementation, a larger LO signal is normally used in conjunction with small core devices to achieve complete changeover with the lower device transconductance, typically around 1 Vp-p.

Since in almost all modern architectures the mixer will be driven from a quadrature digital divider circuit, the normal logic output levels are often compatible with the mixer core. For a bipolar mixer core, the LO divider will typically be CML with a logic swing around 400 mV p-p and simple emitter followers are normally sufficient to buffer the outputs. In CMOS technology the divider architecture may be CML or some other form. Consistent with the switching requirements of a mixer, MOS CML works with larger logic swings, closer to 1 Vp-p. With lower device transconductances, faster edges can require more sophisticated buffering, for example using cross-coupled actively driven current sources to enhance negative going edges.

6.10 **Active mixer enhancements**

As in LNA design, to achieve the desired linearity in the transconductor stage, it may be necessary to use a relatively large bias current. This can be a problem in two respects. Firstly, the IR drop across the load resistors limits the available range at the output and so the compression point. Secondly, a larger current in the switching core increases the device noise during the transition intervals. One method to sidestep this problem is to use a current bleed path from the top of the transconductor to the positive supply.

With a bipolar switching core, a simple resistive path is best (Figure 6.19), as the small signal impedance looking into the switching core is relatively low, and loss of signal is not very significant. For a MOS implementation this tactic is particularly attractive as it would reduce 1/f noise [20] and reduce the transition interval time, but it is not so straightforward in practice, as a P-channel current source will add further noise, while a resistive path is likely to have an impedance comparable with the $1/g_m$ values in the switching core, so that some signal will be lost to VDD. One way around this is to use a dynamic circuit to inject current only during the crossover period. The diversion of current reduces the noise in the switching devices, and the noise due to the diversion source can be made common mode [21].

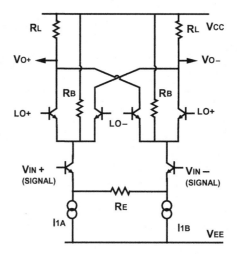

FIGURE 6.19

Use of current bleed resistors in an active mixer to reduce noise and headroom restrictions in the switching core while maintaining noise and linearity of the g_m stage.

6.11 Active quadrature mixers

In a manner somewhat comparable with passive mixers, noise advantages can be achieved by modifying the operation of two active mixers when configured to split a signal into I and Q paths. Due to the phasing of the LO signals, when one path is fully switched, and in the state where the switching core contributes little noise, the other path will be in the transition region. We can observe that for a mixer core in the fully switched region, the voltage at the tail of the differential pairs is pulled higher by the LO peak than the tail of the core in the transition region. This effect can be used to divert current automatically between the I and the Q branches such that each only receives the signal and the bias current when fully switched, and has little current when in transition, with a consequent reduction in the noise from the switching core.

An additional benefit is that the signal from the transconductor is not split between the I and Q paths, but rather is steered to the path that is fully switched, leading to a further improvement in noise, in a manner directly comparable with the 25% passive mixer arrangement.

This scheme works best in bipolar technology where the high device transconductance leads to almost complete steering of the current with normal LO amplitudes [22]. In a practical implementation, small resistors are inserted in the tail connections to aid smooth current division (Figure 6.20). One negative consequence is that each path has a common mode output signal at twice the LO frequency, but in a receiver down-conversion application this is easily rejected and filtered out.

In a CMOS technology, a similar result may be obtained by the use of an additional layer of steering switches, driven by signals at twice the nominal LO frequency, with a $\pi/4$ phase offset (see Figure 6.21).

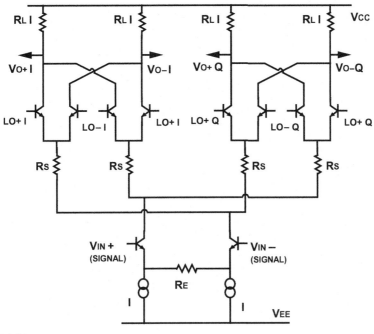

FIGURE 6.20

Quadrature active mixer with self-steering of bias and signal current away from switching core while in changeover state to reduce noise. This also avoids sharing the signal during the periods when each path is fully switched.

In the example shown, the transconductor is realized as a pseudo-differential stage with no degeneration. Improvements in conversion gain of 2 dB are reported in [23] with close-in 1/f noise reduced by 9.5 dB. It is also possible to combine the switching functions by means of 25% LO waveforms at one switch level [24].

6.12 Harmonic rejection mixers

The product terms around the harmonics of the LO signal are usually quite easy to filter out. However, in receivers with a wide tuning range, such as in TV or spectrum-sensing applications, there is a risk of mixing an interferer at the upper part of the input range into the IF band due to (usually odd) harmonics of the LO when the desired signal input is at the lower part of the range. Take the example of a zero-IF receiver with a tuning range from f1 to f2, where f2 is more than three times f1. When the LO is set to f1, then input signals around f1 are mixed into the bandwidth of the channel filters as required. However, input signals around f2 are also able to pass through any input filtering and can mix with the third harmonic of the LO at f1 (Figure 6.22).

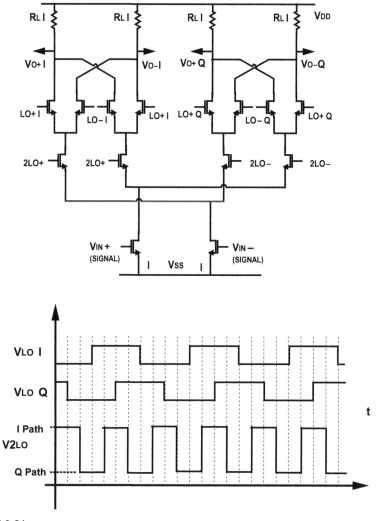

FIGURE 6.21

Quadrature active MOS mixer with steering of bias and signal current away from switching core while in changeover state using additional switch levels.

Tracking filters at the input can help with this, but are not favored due to the inherent nonlinearity of varactor tuned circuits. The two main methods to deal with this problem are (i) to restrict the input signal range with switchable fixed filters to roughly octave spans, or (ii) to modify the mixer structure so as to give a negligible response at the offending harmonics of the LO. Clearly, an ideal multiplier with a

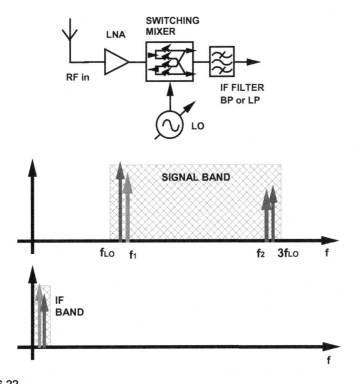

FIGURE 6.22

Spurious responses due to LO harmonic responses in a receiver with wide tuning range.

sinusoidal LO input would meet this requirement but would incur a large noise penalty as explained previously.

A practical strategy is to use a combination of switching mixers (Figure 6.23) where the LO phasing and the input signals are weighted so that the effective summation approximates to the result of a sinusoidal LO, but without the noise problems associated with a true multiplier [25].

Note that with three LO contributions with 45° phase resolution as shown here, it is possible to cancel the third and fifth harmonic responses, but not the seventh, although for the receiver example given, this is unlikely to be an issue. The idea works for both passive and active mixer arrangements in a receiver [25], and can also be used in the transmit path (Figure 6.24) to reduce the filtering requirements [26].

The 45° phase resolution required in the LO phases does, however, add further complexity in the high-speed logic circuitry needed, and hence some power penalty. Rejection of the third harmonic can approach 60 dB.

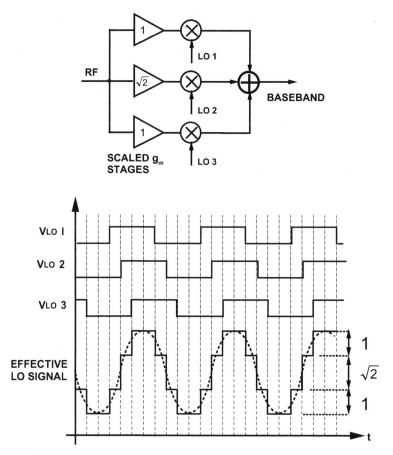

FIGURE 6.23

Harmonic rejection mixer concept with weighted summation of mixer products due to phased LO switching signals.

SUMMARY REMARKS

The two main types of mixers in popular use are current mode passive and current steering active ("Gilbert") types. Both follow a common behavioral model, and are suitable for receive path applications, but the passive type is really limited to down conversion. The passive type has much less 1/f noise in the switches and is attractive for ZIF CMOS designs. In a passive design most noise comes from the terminating transimpedance amplifier, and the switch changeover must avoid creating a transitory high gain loop for the noise. In active mixers, the changeover noise in the switching core is reduced by fast transitions, or by diverting current away during the transition. In complex conversion schemes, both types benefit from effective 25% duty cycle arrangements where the signal is not present in a branch during the changeover.

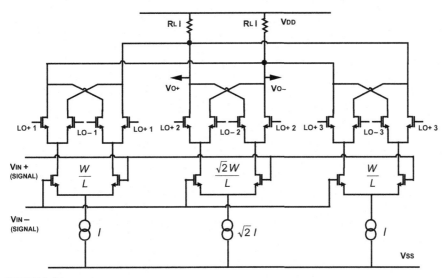

FIGURE 6.24

Implementation of harmonic rejection active mixer.

References

[1] B. Razavi, *RF Microelectronics*, 2nd ed, Pearson, 2012.

[2] A. Abidi, Direct-conversion radio transceivers for digital communications, *IEEE Journal of Solid-State Circuits*, vol. 30, no. 12, December 1995.

[3] B. Razavi, Design considerations for direct-conversion receivers, *IEEE Transactions on Circuits and Systems Part II*, vol. 44, no. 6, June 1997.

[4] K. Kikevas et al., Characterization of IIP2 and DC-offsets in transconductance mixers, *IEEE Transactions on Circuits and Systems Part II*, vol. 48, no. 11, November 2001.

[5] D. Manstretta et al., Second-order intermodulation mechanisms in CMOS downconverters, *IEEE Journal of Solid-State Circuits*, vol. 38, no. 3, March 2003.

[6] A. Mirzaei et al., Analysis and optimization of current-driven passive mixers in narrowband direct-conversion receivers, *IEEE Journal of Solid-State Circuits*, vol. 44, no. 10, October 2009.

[7] W. Redman-White and D. Leenaerts, 1/f noise in passive CMOS mixers for low and zero IF integrated receivers, in *Proceedings ESSCIRC*, September 2001.

[8] S. Chehrazi, Noise in current-commutating passive FET mixers, *IEEE Transactions on Circuits and Systems I*, vol. 57, no. 2, February 2010.

[9] D. Kaczman et al., A single-chip 10-Band WCDMA/HSDPA 4-band GSM/EDGE SAW-less CMOS receiver with DigRF 3G interface and +90 dBm IIP2, *IEEE Journal of Solid-State Circuits*, vol. 44, no. 3, March 2009.

[10] A. Mirzaei et al., Analysis and optimization of direct-conversion receivers with 25% duty-cycle current-driven passive mixers, *IEEE Transactions on Circuits and Systems II*, vol. 57, no. 9, September 2010.

[11] D. van Graas, The fourth method: generating and detecting SSB signals, *QEX*, September 1990.

[12] D. Tayloe, Product detector and method therefore. US Patent 6230000, May 2001.

[13] H. Khatri et al., A SAW-less CDMA receiver front-end with single-ended LNA and single-balanced mixer with 25% duty-cycle LO in 65 nm CMOS, in *Proceedings IEEE RFIC Symposium*, June 2009.

[14] K. Fong and R. Meyer, Monolithic RF active mixer design, *IEEE TRCAS-II*, vol. 46, no. 3, March 1999.

[15] B. Floyd et al., WCDMA direct-conversion receiver front-end comparison in RF-CMOS and SiGe BiCMOS, *IEEE Transactions on Microwave Theory and Techniques*, vol. 53, no. 4, April 2005.

[16] K. Kivekas et al., Calibration techniques of active BiCMOS mixers, *IEEE Journal of Solid-State Circuits*, vol. 37, no. 6, June 2002.

[17] M. Terrovitis and R. Meyer, Intermodulation distortion in current-commutating CMOS mixers, *IEEE Journal of Solid-State Circuits*, vol. 35, no. 10, October 2000.

[18] S. Chen et al., A 65nm CMOS low-noise direct-conversion transmitter with carrier leakage calibration for low-band EDGE application, in *Proceedings IEEE RFIC Symposium*, June 2009.

[19] C. Hull and R. Meyer, A systematic approach to the analysis of noise in mixers, *IEEE Transactions Circuits and Systems I*, vol. 40, no. 12, December 1993.

[20] H. Darabi and A. Abidi, Noise in RF-CMOS mixers: A simple physical model, *IEEE Journal of Solid-State Circuits*, vol. 35, no. 1, January 2000.

[21] H. Darabi and J. Chiu, A noise cancellation technique in active RF-CMOS mixers, *IEEE Journal of Solid-State Circuits*, vol. 40, no. 12, December 2005.

[22] D. Brunel et al., A highly integrated 0.25μm BiCMOS chipset for 3G UMTS/WCDMA handset RF sub-system, in *Proceedings of the IEEE RFIC Symposium*, June 2002.

[23] R. Pullela et al., Low flicker-noise quadrature mixer topology, in *Proceedings of the IEEE ISSCC*, February 2006.

[24] S. Blaakmeer et al., The BLIXER, a wideband balun-LNA-I/Q-mixer topology, *IEEE Journal of Solid-State Circuits*, vol. 43, no. 12, December 2008.

[25] Z. Ru et al., Digitally enhanced software-defined radio receiver robust to out-of-band interference, *IEEE Journal of Solid-State Circuits*, vol. 44, no. 12, December 2009.

[26] J. Weldon et al., A 1.75-GHz Highly Integrated Narrow-Band CMOS Transmitter with Harmonic-Rejection Mixers, *IEEE Journal of Solid-State Circuits*, vol. 36, no. 12, pp. 2003–2015, December 2001.

Integrated Satellite Low Noise Block Down-Converter

7

Pascal Philippe[a], Louis Praamsma[b], and Marcel Geurts[b]

[a]*NXP Semiconductors 2, Esplanade Anton Philips, Campus EffiScience Colombelles BP20000, 14906 CAEN Cedex 9, France*

[b]*NXP Semiconductors, Gerstweg 2, 6534 AE NIJMEGEN, The Netherlands*

INTRODUCTION

Television broadcasting under digital broadcast satellite (DBS) regulations is among the most popular consumer applications operated in Ku-band, generally in the 10.7–12.75 GHz frequency range. The first section of the satellite-TV reception chain consists of the low noise block (LNB) installed in the focus of the outdoor antenna dish. The LNB is in charge of converting the microwave signal down to intermediate frequency in the 950–2150 MHz frequency range.

Many existing LNBs have relatively still low integration level. Their local oscillator (LO) section is frequently implemented using dielectric resonator oscillators (DROs), even though integrated LO generation solutions based on phase-locked-loop (PLL) frequency synthesizers have existed for a few years [1,2]. The signal path is also generally composed of discrete components or monolithic microwave integrated circuits (MMIC) with relatively low integration level.

Nevertheless, integration can offer many advantages to LNB manufacturers. In addition to faster development, integration gives access to the benefits of PLL frequency synthesis techniques. Unlike DROs, no manual frequency alignment is required with a PLL. It is also possible to simplify the LNB housing without compromising the long-term stability of the carrier frequency. But the difficulty of achieving comparable performance to discrete solutions at a reasonable cost has limited the penetration of integrated solutions into the DBS market.

In this chapter, we present the first 10.7–12.75 GHz integrated down-converter with LO synthesizer that has widely penetrated the DBS market. Fabricated in SiGe:C BiCMOS technology, the integrated circuit (IC) achieves state-of-the-art performance for a total current consumption of only 52 mA. Section 7.1 introduces the most important system aspects and explains how the downconverter and LO synthesizer specifications were derived from the LNB requirements. Section 7.2 presents briefly the IC technology and its key features enabling successful integration of high-performance/low-power RF functions. Sections 7.3 and 7.4 address, respectively, downconverter and LO synthesizer design, going from architectural

Manganaro: Advances in Analog and RF IC Design. http://dx.doi.org/10.1016/B978-0-12-398326-8.00007-8

considerations down to circuit-level details. Emphasis is put on the choices made for minimizing cost and power consumption without compromising high-volume manufacturability. Section 7.5 presents the main experimental results of the down-converter IC, revealing a measured performance in line with targeted specifications. Finally, Section 7.6 presents briefly the LNB reference design built around the downconverter/PLL IC.

7.1 System

This section addresses the requirements from a system point of view. Only the "universal" LNBs used in Europe, South America, Africa, and Asia are discussed. Systems used in other regions have similar requirements but differ in details.

7.1.1 Configuration

To make consumer satellite-television reception work, the industry has standardized a configuration that contains an outdoor unit (or the LNB) and an indoor unit (or set top box), connected via a low-cost coaxial cable. The resulting subsystems are driven by the overall system requirements. They can be listed as follows:

1. *Dish antenna:* The satellite transmitted power is very weak: power flux of $-110\,\mathrm{dBW/m^2}$ per transponder is typical [3]. To receive enough power at the receiver, the antenna needs to concentrate the received power flux. The lowest- cost realization is a dish.

2. *K-to-L band converter:* The satellite transmits at Ku-band (10.7–12.75 GHz). The antenna is connected via a coax cable to the indoor unit. Low-cost coax cables show strong attenuation at Ku-band frequencies. Therefore the signals from the antenna are converted from Ku band down to 950–2150 MHz (L band). This conversion requires the following elements in the outdoor unit:

   a. *Mixer:* A mixer at Ku band is rather simple, since the IF band and Ku band have a large relative frequency distance. Single transistors (such as pseudomorphic high electron mobility transistors (pHEMT) and bipolar junction transistors (BJT)) or diode mixers are deployed for this function. Unfortunately the mixers have a rather a large noise figure, therefore low noise amplification is required before mixing in order to meet the system requirements.

   b. *Bandpass filter:* Since the simplest mixers are double-sideband mixers, image rejection must be implemented. This is typically achieved using a distributed filter in the so-called hairpin configuration.

   c. *Local oscillator:* The phase noise of the LO must be taken into account in the available phase noise budget of the link. Realizations use BJT oscillators with DROs or PLL frequency synthesizers as will be discussed later. The DROs have a very low phase noise, but need to be aligned in production and suffer from frequency drift over their lifetime ("aging"). PLLs, which use a crystal

(XTAL) as a reference, do not require alignment in production and have superior behavior in terms of aging.

d. *Low noise amplifier at Ku band:* To mask the noise of the mixer, the signals from the antenna must be amplified first. As amplification at Ku is more expensive than at L band, the number of amplification stages is reduced to the minimum. A typical configuration uses two-stage amplifiers. Often pHEMTs are used, but nowadays BJTs are also possible. Discrete transistors are chosen for cost reasons.

3. *Feed horn:* The feed horn pattern should be matched to the dish: if the pattern is larger than the dish, noise and unwanted signals will be received. In the case of a smaller pattern, part of the received signals is not used. The feed should be able to receive the signal under horizontal and vertical polarizations. Because of earth curvature, the polarization can be rotated with respect to local horizontal or vertical references. Therefore the feed must have the possibility to be adjusted around its axis.

7.1.2 LNB block diagram

Figure 7.1 shows the block diagram of a universal single LNB with integrated downconverter and LO synthesizer.

The RF signal received from the horizontal and vertical polarization feeds is amplified using a two-stage low noise amplifier (LNA). As one out of the two possible polarizations is received at a time in a single LNB, the second stage is common

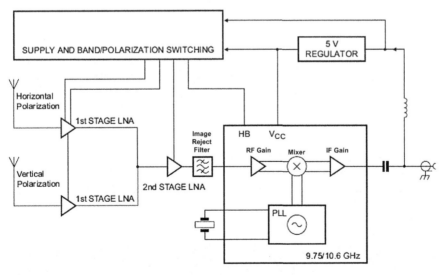

FIGURE 7.1

Low noise block application diagram with integrated downconverter.

to both paths. The interface between the first and second stage is designed such that if one first stage is off, the output impedance of this stage is transformed into an open for minimizing loss. The LNA is followed by the bandpass filter in charge of image rejection, which is typically implemented as a hairpin filter as mentioned earlier. After the filter, the integrated downconverter translates the RF signal down to intermediate frequency (IF) in L band (950–2150 MHz) by mixing with the LO signal generated on-chip. At the IF output, a bias-T separates the IF output signal from DC and low-frequency control signals carried from the set top box (STB) up to the LNB via the coaxial cable. The 5 V supply voltage for the downconverter and the LNA is generated out of the control signal coming from the STB. The control signal includes a two-level DC voltage (13/18 V typical) for selecting between vertical and horizontal polarizations, on top of which a 22 kHz tone is added or not for selecting between high band or low band, respectively.

7.1.3 From recommendations toward product specifications

In this section, the conversion of LNB recommendations into specifications for the downconverter and LO synthesizer functions of the LNB is discussed. The technical recommendations by SES Astra [4] for a universal single LNB presented in Table 7.1 are taken as a reference.

7.1.3.1 LO synthesizer requirements

The typical LO frequencies are system given: 9.75 GHz for low band and 10.6 GHz for high band. Also the deviation is stated: ± 5 MHz. For a PLL-based LO generator, the relative frequency stability is identical to that of the reference frequency. The worst-case situation occurs at 10,600 MHz LO frequency, where the allowed stability of ± 5 MHz corresponds to ± 472 ppm frequency error. This sets the requirement for the reference oscillator.

Concerning LO phase noise, the recommendation assumes declining phase noise with increasing offset frequency. This is a good assumption for a free-running oscillator. However, for a PLL system, the phase noise follows a different law with frequency offset. As the spectral distribution of phase noise is not of major importance in the system, it is not necessary to fulfill the recommended spectral mask in each point. A more relevant specification is to consider the phase noise power integrated over the system bandwidth. The lower integration limit is determined by the bandwidth of the carrier recovery loop in the receiver. Low frequency variations of the carrier are corrected up to this value. A value of 10 kHz is appropriate for Digital Video Broadcasting (DVB) systems. The upper integration limit is determined by the signal bandwidth. The channel bandwidth is 26 MHz, leading to an upper integration limit of 13 MHz single-side band. Integration of the phase noise spectral density over 10 kHz–13 MHz sets the phase noise specification to $1.7°$ RMS.

A reference frequency of 25 MHz was chosen for the PLL, although a better option for minimizing PLL noise (see section on PLL design) would have been to take the highest possible reference frequency compatible with integer division of the

Table 7.1 Astra Technical Recommendations for Universal Single LNB

No.	Parameter			Value			Unit
				Minimum	Typical	Maximum	
1	Input frequency	Low band		10.70–11.70			GHz
		High band		11.70–12.75			GHz
2	Output frequency	Low band		950–1950			MHz
		High band		1100–2150			MHz
3	Local oscillator frequency	Low band		9.745	9.750	9.755	GHz
		High band		10.595	10.600	10.605	GHz
4	Phase noise	Low band and high band	At 1 kHz offset			−50	dBc/Hz
			At 10 kHz offset			−75	dBc/Hz
			At 100 kHz offset			−95	dBc/Hz
			At 1 MHz offset			−105	dBc/Hz
			At ≥1 MHz offset			−115	dBc/Hz
5	Conversion gain	In 26 MHz bandwidth		50		60	dB
6	Gain ripple	Low band			3	5	dB
		High band			3	5	dB
7	Noise figure	Low band			1.1	1.3	dB
		High band			1.3	1.5	dB
8	Image rejection			40			dB
9	1 dB compression point			0			dBm

Astra Universal Single LNB — Technical Recommendation — V1.0 September 2007

Table 7.1 Astra Technical Recommendations for Universal Single LNB *Continued*

Astra Universal Single LNB—Technical Recommendation—V1.0 September 2007

No.	Parameter		Value			Unit
			Minimum	Typical	Maximum	
10	Third-order intermodulation (2 tones)	Intercept point	10			dBm
11	Output impedance			75		Ω
12	Return loss		8			dB
13	Cross talk (different polarizations)		20			dB
14	In band spurious			−65	−60	dBm
15	LNB supply voltage (control signal)	Vertical polarization (signal Ca)	11.5		14.0	V
		Horizontal polarization (signal Cb)	16.0		19.0	V
16	High band selection (control signal Cc)	Frequency	18	22	26	kHz
		Duty cycle	40	50	60	%
		Peak-to-peak voltage	0.4	0.6	0.8	V
		Transition time	5	10	15	μs
		Load impedance at 22 kHz	70			Ω
17	Current consumption per LNB			100	200	mA

LO carrier frequency, which is 50 MHz. Nevertheless, 25 MHz was preferred because of the lower cost of 25 MHz crystals.

7.1.3.2 Down-converter requirements

The downconverter requirements are determined from a signal budget analysis of the complete LNB, considering the architecture of Figure 7.1.

The typical recommended conversion gain of a LNB is 55 dB. Taking into account LNA gain and bandpass filter loss, the downconverter needs to have about 35 dB conversion gain. Thanks to more than 20 dB gain of low noise amplification, the noise figure (NF) of the downconverter can be relaxed to 7 dB without compromising the NF of the LNB, which has to be very low, typically in the order of 1 dB or less. As the downconverter includes the output stage of the LNB, its 1 dB compression point (P1dB) and output third-order intercept (OIP3) must be at least at the level recommended for the complete LNB, meaning P1dB > 0 dBm and OIP3 > 10 dBm. In fact, 1 dB more is provisioned to account for bias tee loss.

For achieving the recommended 40 dB image rejection, a possible option is to implement most of the rejection in the downconverter. This can be accomplished either using an image-reject mixer or an on-chip filter. However, because of the higher current consumption and/or die cost of such an approach, performing most rejection off chip looks preferable. This can be achieved relatively easily and at low cost with the bandpass hairpin filter printed on the circuit board.

On top of signal distortion and image response, the IF signal at downconverter output can be harmed by spurious responses generated within the downconverter. Those spurious signals can have several origins. Among others, they can be related to harmonics of the reference frequency, high-order mixing products, or insufficient rejection of the 22 kHz band selection tone present in the system.

7.1.3.3 Current budget

The typical supply voltage of LNBs is 5 V, which is generated out of the 13/18 V polarization control signal using a linear regulator. Despite the fact that this approach wasting a lot of power in the regulator, it is preferred to a switched mode DC/DC converter solution for its low cost. To minimize the power dissipated into the regulator, a rather low supply current is therefore required. Discrete LNB realizations already consume far less than the 200 mA maximum recommendation of Astra. Therefore a budget of 60 mA maximum was proposed for the downconverter in order to come up with an integrated solution consuming no more than discrete implementations.

7.1.3.4 Summary of downconverter and LO synthesizer specifications

Table 7.2 summarizes the main specifications of the integrated downconverter and LO synthesizer. Four different conversion gains are provisioned to accommodate different front-end implementations.

Table 7.2 Specifications of Integrated Downconverter with LO-PLLs

Description	Conditions	Minimum	Typical	Maximum	Unit
Current consumption				60	mA
RF input band	Low band	10.70		11.70	GHz
	High band	11.70		12.75	GHz
IF output band	Low band	950		1950	MHz
	High band	1100		2150	MHz
Input return loss	50Ω reference impedance			−10	dB
Output return loss	75Ω reference impedance			−10	dB
Conversion gain	Low gain		35		dB
	Medium-low		38		dB
	Medium-high		42		dB
	High gain		45		dB
SSB noise figure	Including image-reject filter			7	dB
1 dB output compression point		1			dBm
Third-order output intercept point		11			dBm
Spurious output signals	In presence of signal; $P_{out} \geq -30$ dBm			−40	dBc
	In absence of signal			−60	dBm
LO leakage at RF input				−35	dBm
LO frequency	Low band	9.75			GHz
	High band	10.60			GHZ
Reference frequency		25			MHz
Integrated LO phase noise				1.7	°RMS

7.2 **IC technology**

A downconverter with LO synthesizer IC has been designed in NXP's QUBiC4X BiCMOS technology [5]. The technology features a rich set of active and passive components enabling the integration of complex, high-performance, and low-power RF functions at low cost. The key component consists of a SiGe:C NPN hetero-junction bipolar transistor (HBT) with 110/180 GHz fT/fMAX, respectively. The HBT features an inside L-shaped spacer between base and emitter which creates sub-lithographic emitter openings ($0.4 \times 1\,\mu m^2$ minimum drawn emitter size). This L-shaped spacer also reduces emitter plugging, eliminates dry etching of the base mono-silicon, and improves reliability. The NPN is fabricated with a monocrystalline emitter. Collector-to-substrate and collector-to-base capacitance have been reduced, respectively, by the use of deep trench isolation and shallow trench isolation. Together with high-performance HBT, the QUBiC4X process offers $0.25\,\mu m$ CMOS transistors enabling implementation of digital processing and control functions for the realization of advanced mixed-mode circuits.

The QUBiC4X technology also includes high-performance passive components that allow the exploitation of the high-performance capability of the HBT in all kinds of RF functions as commonly found in RF receivers, transmitters, or frequency synthesizers. They consist of a wide variety of resistors, MOS, and junction varactors with high-quality factor and MIM capacitors with high density of $5\,fF/\mu m^2$. Device interconnect is facilitated thanks to five metallization layers. The two top metal layers are $2\,\mu m$ and $3\,\mu m$ thick, enabling integration of high-quality inductors and low-loss transmission lines. Finally, substrate loss and crosstalk are minimized by use of a high-resistivity silicon substrate.

7.3 **Downconverter design**

7.3.1 **Architecture**

The downconverter is implemented as a single-frequency conversion receiver, converting the RF signal received in Ku band (10.7–12.75 GHz) down to IF in L-band (0.95–2.15 GHz) (Figure 7.2). It consists of the following functions:

- *The low noise amplifier (LNA):* It provides sensitivity and also performs single-ended to differential conversion. Its key parameters are gain, noise figure, and input matching.
- *The RF Filter:* It rejects out-of-band signals, in particular the image band, which is converted to the desired IF frequency signal.
- *The mixer:* It performs frequency conversion of the RF band down to IF by mixing with the LO signal generated by the on-chip frequency synthesizer.
- *The IF amplifier:* It amplifies the signal up to the desired output power, and performs differential to single-ended conversion. It is also used for compensating

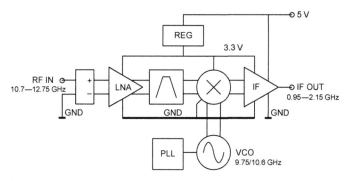

FIGURE 7.2

Downconverter block diagram.

frequency-dependent loss of the IF coaxial cable. The important parameters of the IF amplifier are gain, gain shape, linearity, and output impedance.

7.3.2 General design considerations

The current consumption of the output stage is dictated by the power to be delivered to the output. At up to 6 dBm output power (P_{out}), a peak current (I_{peak}) of 10 mA resulting in 750 mV voltage swing must be delivered into the load (R_{load}), since:

$$P_{\text{out}} = 0.5 \cdot I_{\text{peak}} \cdot R_{\text{load}}. \tag{7.1}$$

For the intermediate stages, the current is determined by linearity and bandwidth requirements. High linearity requires strong emitter degeneration. The input voltage of third-order intercept point (IIP3) for a degenerated bipolar stage [6] is indeed:

$$V_{\text{IIP3}} = 4V_t \left(1 + \frac{R_e I}{V_t}\right)^{3/2}, \tag{7.2}$$

where R_e is the emitter degeneration resistance, I is the DC current, and V_t is the thermal voltage.

In addition, the output bandwidth sets a limit to the load resistor, R_c, placed in the collector of the amplifier. It follows that if a certain low-frequency voltage gain, G_v, is desired, the emitter resistance is defined, since we have:

$$G_v = \frac{R_c}{R_e}. \tag{7.3}$$

In consequence, Eq. (7.2) shows that the only parameter left for achieving the desired linearity is the DC current. If ever the current cannot be increased up to the desired level because of voltage headroom issues, then the gain must be reduced by increasing R_e. From this simple analysis, we can conclude that minimizing the current consumption requires high gain in the last stages of the downconverter.

With regard to noise behavior, the conclusion is the opposite. A high gain is required in the input stage for masking the noise added by the following stages. The best trade-off between these conflicting requirements consists of putting just enough gain in the RF input stage for masking sufficiently the noise added by the rest of the chain.

In spite of the single-ended input and output, the internal architecture of the downconverter is fully differential. The reason is that differential circuits do not inject signal into ground or supply terminals. Coupling between stages is thus prevented. In addition, differential design allows accurate control of the impedances placed in the emitter and collector of the transistors, since they are not dependent on impedances of ground and supply lines, which are often difficult to quantify precisely.

7.3.3 Stability considerations

It is well known by design engineers that amplifiers with high gain can become unstable, if for some reason a too-large portion of the output power is fed back to the input. However, the risk of instability is not limited to amplifiers and is also a major concern in the design of a downconverter. Like in an amplifier, a first risk factor is related to the high conversion gain, targeted over 40 dB. Another risk is associated with the single-ended architecture of the RF input and IF output stages, which was chosen for minimizing the number of pins. The reason why the latter represents a risk is due to lower rejection of common ground signals compared to a full differential architecture, which finally results in lower isolation between input and output.

The stability of a frequency converter can be examined using a feedback loop model as in the case of an amplifier, except that we have now to consider mixing with the LO signal. Quite often, the study is limited to the RF frequency equal to half LO, which is a particular situation that facilitates the analysis because the IF frequency is also at half LO, due to the fact that:

$$f_{IF} = f_{LO} - f_{RF}. \tag{7.4}$$

With equal input and output frequencies, we are back to the easy case of an amplifier, which can be addressed using the well-known theory of feedback loop amplifiers. Most probably for this reason, there is sometimes the belief that instability in a frequency converter can only occur at RF frequency equal to half LO. However, this is not the only possibility, but finding the others requires considering more mixing products, and not limiting the analysis to the mixing term of order 1 in LO, as in Eq. (7.4).

A general approach to this problem is given in [7], but it is generally sufficient to consider only the dominant mixing products. They are of order 1 and −1, where "order" refers to the multiplication factor of the LO frequency involved in the relationship between the output (IF) and input (RF) frequencies. In our downconverter wherein the wanted RF is above LO, the IF frequency is given by the mixing term of order −1:

$$f_{IF} = f_{RF} - f_{LO}. \tag{7.5}$$

We define $G(f_{RF})$ the conversion gain associated to this product and adopt the convention to take the input frequency as argument for all transfer functions that will be further defined. As instability implies feedback, we assume that the output signal can be coupled back to input in some way, with transfer function $H(f_{IF})$. We admit that the backward transfer function is dominated by radiated or conducted coupling and therefore that no frequency conversion is involved. At this stage of the analysis, we see that the signal is back to input at a different frequency after one turn, since the frequency is f_{IF} instead of f_{RF}. As a consequence, no conclusion can be drawn yet about stability since this implies the examination of a loop transfer function with the same input and output frequencies. However, the condition of equal input and output frequencies can be fulfilled if we consider a second turn of the signal through the loop. Indeed, the downconverter does respond to an input signal at f_{IF} with conversion gain $G(f_{IF})$. This conversion is associated with the mixing term of order 1 and gives an output frequency at f_{RF}:

$$f_{RF} = f_{LO} - f_{IF}. \tag{7.6}$$

We admit again that the output signal can be coupled back to input without frequency conversion, with reverse gain $H(f_{RF})$. Now, we see that the signal is back to input with the same frequency after two turns. It follows that the stability can be analyzed using well-known theory developed for feedback amplifiers provided one takes as loop gain:

$$G_{loop}(f_{RF}) = G(f_{RF}) H(f_{IF}) G(f_{IF}) H(f_{RF}), \tag{7.7}$$

which is actually the product of the two loop gains associated with down-conversion and up-conversion loops.

The application of this theory to our downconverter can be carried out using the behavioral model shown in Figure 7.3. Due to the differential architecture, we consider that only the single-ended input and output stages have a reference to ground. Although other coupling paths are possible, we admit that the dominant feedback path is caused by the common inductance in the ground path to the printed circuit

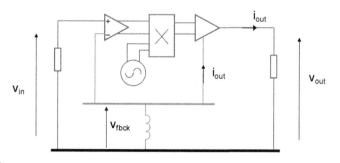

FIGURE 7.3

Behavioral downconverter model for stability analysis.

board. Then, assuming ideal decoupling of the supply pin of the single-ended IF output stage, it follows that the signal current delivered into the load also flows into its ground line. This IF current is converted to voltage by the common ground inductance. As the low noise amplifier is actually a differential amplifier having one of its terminals internally grounded, the ground voltage is perceived as a feedback input signal. Using this model, the loop gain as defined in Eq. (7.7) is given by:

$$G_{\text{loop}}\left(f_{\text{RF}}\right) = Y_c\left(f_{\text{RF}}\right) j L_g \omega_{\text{IF}} Y_c\left(f_{\text{IF}}\right) j L_g \omega_{\text{RF}}, \tag{7.8}$$

where $Y_c(f)$ is the conversion transadmittance of the downconverter in the forward path. To avoid any risk of instability, the loop gain must verify:

$$\left|G_{\text{loop}}(f)\right| < 1. \tag{7.9}$$

Fulfilling this requirement imposes conditions on ground inductance and out-of-band conversion gain. For instance, assuming a ground inductance $L_g = 100$ pH, the conversion transadmittance at half LO frequency (5 GHz) needs to verify:

$$\left|Y_c\left(f_{\text{LO}}/2\right)\right| < 0.32\text{S}. \tag{7.10}$$

Converting this result into voltage gain in dB in a 50 Ω system gives:

$$\left|G\left(f_{\text{LO}}/2\right)\right| < 24 \text{ dB}.$$

As the in-band gain needs to be over 40 dB, we conclude from this analysis that a relative attenuation of more than 16 dB is necessary at $f_{\text{LO}}/2$. This result highlights clearly the importance of minimizing the common ground inductance and of rejecting out-of-band signals as much as possible.

7.3.4 Low-noise amplifier

7.3.4.1 RF package interface

The RF signal is fed into the device using a 50 Ω microstrip line, whereas the input of the packaged device itself resembles a coplanar wave guide with its RF input enclosed by ground pins. This pseudo coplanar wave guide is continued inside the package via the bonding wires connecting the die to the package leadframe. It appears that this representation of the package parasitics by a coplanar transmission line with impedance close to 50 Ω is valid up to the input frequencies of the downconverter.

7.3.4.2 RF gain stage

The functions of the RF gain stage are to provide low noise amplification in combination with single-ended to differential conversion, with an input impedance of around 50 Ω. The RF gain stage is followed by an image filter supporting high output impedance. The requirement for input IP3 of the LNA is moderate (+6 dBm − 36 dB = −30 dBm).

Following a classical approach, the RF gain stage implements inductive degeneration for achieving low noise figure. Its input impedance scales with $L_e\omega_T$, where ω_T is the transition frequency, and L_e is the emitter degeneration inductance. As ω_T is a function of I_c, the current consumption will be given by the amount of emitter inductance that can be tolerated to achieve the gain versus the collector current that is needed to achieve high ω_T for the given transistor size.

Then, the dimensioning of the low noise amplifier follows the same approach as the familiar optimization of a single-ended LNA, taking advantage of the virtual ground of the differential pair which would be a package parasitic in a single-ended LNA. The differential LNA is a straightforward combination of two single-ended LNAs, with $100\,\Omega$ input impedance.

The resulting differential input of the LNA can be converted to a single-ended one as described in [8]. A two-turn inductor in parallel with the input ensures single-ended-to-differential conversion, allowing, at the same time, matching to $50\,\Omega$ by proper dimensioning of the input capacitors. The amplifier with integrated balun is shown in Figure 7.5 together with the mixer and filter.

7.3.4.3 Band filter

The image filter is implemented between the LNA and the mixer, both having a relatively high impedance. The filter should have a sharp roll-off for image suppression, while at the same time having a flat in-band response. An elegant implementation of such a filter consists of two weakly coupled LC resonant circuits.

Figure 7.4 below shows the transfer characteristic of coupled LC filters. A flat in-band gain response is obtained when the product of the coupling coefficient, k, and the quality factor, Q, of the resonators is equal to one:

$$k \cdot Q = 1, \tag{7.11}$$

while the relative 3 dB bandwidth equals:

$$BW_{\text{rel}} = 1.5/Q. \tag{7.12}$$

For reference, the bandwidth of a single tuned filter is only $1/Q$.

The bandwidth is given by the system requirement from 10.7 to 12.7 GHz or 17%, leading to a required Q of around 8, and thus a coupling coefficient of 12%. The quality factor is compatible with the thick metal BiCMOS process used.

A transformer layout of the two inductors would lead to couplings of more than 50%, while the magnetic coupling factor between two inductors placed next to each other is not sufficient, $\sim$1% or less, and therefore the final coupling between the stages has to be defined by small capacitors between the resonators, and they can be used to fine-tune the shape of the frequency response depending on the realized Q of the resonators.

The quality factor of the tank will be a combination of the impedances of the LNA and the mixer and the inductor of the resonator. With input and output values of

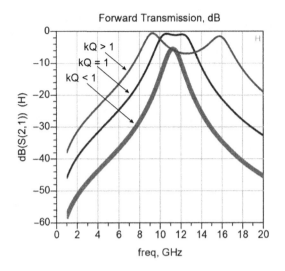

Forward Transmission, dB

kQ > 1
kQ = 1
kQ < 1

FIGURE 7.4

Band filter characteristic depending on coupling factor.

500 Ω, the inductor will be around 1 nH, resonating with 0.4 pF to 12 GHz, and the final component values can be optimized using a circuit simulator.

7.3.5 **Mixer**

The mixer is a standard double-balanced Gilbert cell. Apart from the full differential operation, the advantage of this mixer architecture is high isolation of the IF port from the RF and LO ports. As a benefit, the contribution of the LO path to the IF noise is negligible compared to the downconverted RF noise.

In the mixer, degeneration resistors are used for two reasons. The first one is for stabilizing the gain and the input impedance against transconductance variations with temperature. The gain stabilization is further improved by choosing similar resistor types in the emitter and the collector. The second reason is for improving the linearity. The output IP3 requirement translates into an input IP3 better than −5 dBm at mixer input. On an impedance level around 500 Ω, this results in a voltage level of 500 mV. Such a large level makes some degeneration in the mixer necessary, especially because the mixer should give a negligible degradation of the overall IP3. The total lineup of LNA, band filter, and mixer is shown in Figure 7.5.

7.3.6 **IF amplifier**

The last section of the downconverter is the IF amplifier, composed of two stages. The first stage sets the level and the shape of the voltage gain, while the second stage provides impedance transformation and differential to single-ended conversion.

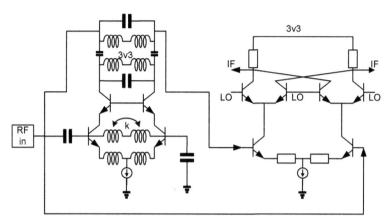

FIGURE 7.5

Simplified schematic of LNA/mixer with integrated band filter.

The output stage shown in Figure 7.6 uses a totem-pole structure. Thanks to the differential drive of the bottom and top transistors and the proper choice of resistor values ($R_{e1} = R_{e2} = R_L$), the top transistor operates at a constant current, presenting infinite small-signal impedance to the bottom transistor. As a consequence, all but the collector signal current from the bottom transistor is delivered into the load in small-signal conditions. However, the output impedance is set by the degenerated top transistor that behaves as an emitter follower. The output impedance can therefore be made very low.

Under large-signal conditions, the top transistors start supplying some current to the load, making the stage work as a push–pull or class AB amplifier. As a benefit,

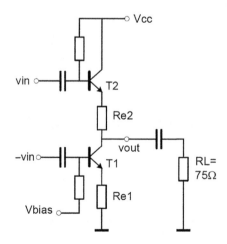

FIGURE 7.6

Simplified schematic of IF amplifier output stage.

the compression point can be achieved at lower DC current than with a pure class-A output stage.

However, this totem-pole structure requires a rather large supply voltage for stacking two transistors and two resistors carrying large current, also for accommodating the large voltage swing at its differential inputs, which is equal to the output swing because of unity voltage gain. Fortunately, this drawback could be easily overcome by using the un-regulated 5V power supply present in the LNB for supplying the IF amplifier, thus leaving enough headroom for all devices.

The gain stage that precedes the output stage is a two-stage differential amplifier, consisting of a cascoded common-emitter amplifier with high voltage gain, followed by a common-collector stage providing low drive impedance to the output stage, as shown in Figure 7.7. The voltage gain of the cascoded amplifier is determined by the ratio of the collector to emitter impedances. The collector resistance and collector current are set as function of the bandwidth requirement and maximum voltage swing anticipated at the collectors. The emitter impedance is then derived from the desired voltage gain. It is not just a resistor but a mixed parallel/series RC network of decreasing impedance with increasing IF frequency. As a result, the gain increases with IF frequency. This behavior is implemented on purpose for compensating increasing losses in the IF coaxial cable.

The downconverter gain variants mentioned earlier are implemented by changing the IF gain. The different gains are generated by changing the emitter degeneration impedance of the differential cascoded amplifier. When decreasing the gain, the input IP3 of the cascoded amplifier is increased. This is favorable for maintaining the same IP3 at the output of the IF amplifier. However, when the IF gain is lowered, a higher output level is required from the mixer for a given IF output power. There is a risk that the linearity could be deteriorated by the mixer. In consequence, the LNA/mixer needs to have sufficient linearity compatible with the lowest IF gain for achieving the same output IP3 for all gain levels.

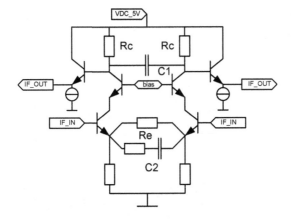

FIGURE 7.7

Schematic of IF amplifier gain stage.

7.4 LO-PLL design

7.4.1 Architecture

The architecture of the LO generator is presented in Figure 7.8. The LO generator consists of a RF Voltage-Controlled Oscillator (VCO), the frequency of which is controlled by an integer-N Phase-Locked Loop (PLL). The LO frequency is switchable between 9.75 and 10.6 GHz, corresponding to low and high band respectively. Thanks to the choice of a 25 MHz reference frequency, the division ratios are even numbers in both bands. They are respectively 390 and 424. The advantage of an even division number is that the first stage of the RF programmable divider can have a fixed frequency division-by-2. A divider-by-2/3 would have been otherwise required, which consumes significantly more power and can potentially create spurious tones in the VCO spectrum because of changing state over a comparison cycle.

While the PLL has rather conventional architecture, it was optimized for low current consumption using innovative circuit design solutions. Since most of its functional blocks are discussed in detail in the next sections, we give here only a brief description of the PLL. The frequency divider, which divides the VCO frequency down to the reference frequency, is based on a cascade of divide-by-2/3 cells. The reference signal is generated from a 25 MHz crystal oscillator implemented on-chip. The phase comparator, which measures the phase error between the reference and divided signals, consists of a phase-frequency detector (PFD), as commonly found in modern PLLs. The task of the chargepump (CP) that follows the PFD is to convert the phase error into charge,

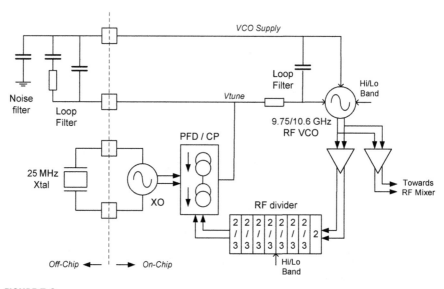

FIGURE 7.8

Architecture of the integrated LO generator.

by generating current pulses of constant height but variable width proportionally to the phase error. These current pulses are converted to control voltage for the VCO by integration in the loop filter capacitance. As capacitors in the order of 1 nF are required, the loop filter is implemented off-chip using RC discrete components. Concerning the VCO, it integrates a LC resonant circuit of high-quality factor for achieving low phase noise. The VCO tuning range is divided in two sub-bands centered on low- and high-band LO carrier frequencies for having about the same tuning voltage in both bands.

7.4.2 Implementation strategy

Most functional blocks have differential circuit architecture but some like the PFD and the charge-pump have a single-ended structure. Differential structures are generally preferred in integrated circuits because of their interesting properties in both situations of aggressor or victim. As victim, differential circuits have the ability to reject common mode parasitic signals eventually present on their ground and supply lines. As aggressor, differential circuits present the advantage of very low current signals injection into ground and supply lines, thanks to the constant current mode of operation. As a result, cohabitation of differential blocks on a shared supply/ground domain is much easier. However, the penalty for benefiting from these interesting properties can be a relatively high current consumption in comparison to other techniques, such as CMOS design. As CMOS circuits consume only during transitions, the consumption of a CMOS equivalent implementation can be much lower, in particular when the signal frequency is low. Following this reasoning, we came to the conclusion that the PFD, which operates at 25 MHz, would be better implemented in CMOS. As a consequence, the charge pump, which has a single-ended output and follows the PFD, was also naturally designed as a single-ended circuit. Although not a CMOS design, we followed a switched-mode design approach resulting in no static current consumption in the charge pump. To limit the risks of excessive interference between blocks, a differential approach was kept for all other functions, although some other low-frequency ones could have been implemented in CMOS as well for a higher reduction of the power consumption.

In addition to the proper balance between differential and single-ended circuits, the success of an integrated solution also depends on other precautions taken for minimizing interferences between blocks sitting on the same die. The interferences can be of different natures, coupled or conducted. Coupling can be inductive or capacitive. Interferences can be conducted via the substrate, block interconnects or ground supply lines. The interferences can be localized on-chip, involve bonding wires, or package leads that connect the die to the outside world. This is a very complex problem, to which there is no simple answer, and for which it is also generally difficult to quantify the impact of the measures taken. Nevertheless, there are general methods which can be adopted for mitigating the risk of interference between blocks. To combat conducted interferences, the method consists generally of attaching aggressor and victims to different supply/ground domains, in order to minimize the impedance of shared ground and supply lines. However, the separation into different domains is limited by the number of pads or terminals available for grounding or supply. In our downconverter, four domains were defined for the PLL. One domain is devoted to the VCO, which is a very sensitive function susceptible to being easily modulated by all kinds of parasitic signals generated by surrounding blocks. Another sensitive block is the XO, which also runs under a specific supply/ground domain. A third domain is attributed to the PFD and charge-pump, considered as potential aggressors because of their single-ended operation and because of the pulsed nature of the charge-pump output signal. The fourth domain is for the RF frequency divider, which can be considered as an aggressor too, considering that it generates by principle a lot of subharmonics of the VCO signal. In consistency with the definition of different domains, measures were also taken during circuit layout for maximizing the substrate isolation in order to keep the separation between domains as effective as possible.

Each domain has its own regulated supply, generated on-chip by regulators from the external 5 V supply. The primary function of the regulators is to generate a supply voltage around 3 V, adapted to the breakdown voltage of the transistor technology. Another important function of the regulators is to provide additional rejection of the 22 kHz tone, on top of that already given by the 5 V LNB regulator.

7.4.3 Noise budget specification

All noise sources present in the different functional blocks of the synthesizer contribute to LO carrier phase noise at VCO output. For good prediction of phase noise, a budget must be established that specifies the level of all local noise sources, taking into account the transfer function that relates LO carrier phase noise to local noise. The role of budgeting is also to distribute the noise contributions evenly over the different sources, as far as possible, so as to avoid having phase noise dominated by a single contributor.

The common practice is to specify VCO/PLL noise as single sideband voltage noise power relative to carrier power. This ratio is generally noted $L(f_m)$. As sideband voltage noise is dominated by phase noise in a VCO/PLL, the following approximation is generally well verified over the offset frequency range of interest:

$$L(f_m) \approx \frac{1}{2}S_\phi(f_m), \qquad (7.13)$$

where $S_\phi(f_m)$ is power spectral density of phase noise. As $L(f_m)$ is a measure of noise relative to carrier, its level is naturally given in dBc/Hz.

From the noise analysis of a PLL, it can be shown that the PLL phase noise measured at the VCO output can be broken into in-band noise and out-of-band noise [9,10]. The in-band phase noise is mostly determined by noise sources modulating the phase reference of the PLL. In addition to noise sources in the reference oscillator itself, the contributors include noise sources of the reference amplifier, the phase detector, the charge-pump, and the RF divider. In contrast, the out-of-band noise is almost entirely set by the RF oscillator. The crossover frequency that delimits the two noise regimes is determined by the unity gain bandwidth of the PLL loop gain.

It is useful to mention that in a PLL, the in-band portion of phase noise measured at VCO output obeys a general scaling rule in function of the reference frequency, F_{REF}, and VCO frequency, F_{VCO},

$$L(f_m) = \text{FOM} + 20\log(F_{VCO}) - 10\log(F_{REF}) \text{ in dBc/Hz}, \qquad (7.14)$$

where FOM is the factor of merit that characterizes the noise performance of the PLL. One understands from this law the remarks made earlier that starting from a high reference frequency is generally desired for minimizing in-band noise. The above scaling rule is also useful for changing the reference point of phase noise specification. For PLL circuit design, it is often more convenient to give a specification that is referred to the input of the PLL, which operates at the Reference frequency. This is easily accomplished by replacing F_{VCO} by F_{REF} in Eq. (7.14); this is equivalent to subtracting $20\log(N)$ from the figure referenced to VCO output.

Following these considerations, the PLL noise budget given in Table 7.3 can now be more easily understood. The budget was established for 1.7° rms maximum integrated phase noise (10 kHz–13 MHz). The in-band phase noise of the PLL is referred to the 25 MHz input, whereas the VCO is specified at a carrier frequency of 10 GHz. The specification implies a figure of merit of −219 dBc/Hz typical, as defined from Eq. (7.14).

7.4.4 **LO signal source**

The LO source consists of an amplitude-controlled RF VCO with separate buffered outputs towards the RF mixer and divider.

Table 7.3 PLL Noise Budget Specification

Parameter	Conditions	Typical	Maximum	Unit
VCO phase noise at 10 GHz	100 kHz offset	−93	−90	dBc/Hz
PLL noise referenced to 25 MHz input	10–100 kHz offset			
XO/XO buffer contribution		−150	−147	dBc/Hz
RF divider contribution		−150	−147	dBc/Hz
PFD/charge pump contribution		−150	−147	dBc/Hz
Total		−145	−142	dBc/Hz
Crossover frequency		90		kHz

Figure 7.9 shows a simplified circuit diagram of the RF VCO, which is a modified common base differential Colpitts LC oscillator [11]. The common base differential pair can be viewed as the oscillator loop amplifier. Positive feedback is applied by means of capacitive divider, C_1, C_2, for generating negative resistance. At the resonance of the LC circuit, the signal at the collector is fed back in phase with the input signal at the emitter. The feedback ratio is defined as:

$$\beta = \frac{C_2}{C_1 + C_2}. \tag{7.15}$$

An optimum feedback ratio exists that minimizes phase noise, which is $\beta \approx 0.3$ for a standard Colpitts oscillator [12].

The modified Colpitts architecture implements a cross-coupled pair at the bottom of the circuit, which improves significantly the oscillator performance [13]. It brings additional negative resistance, which boosts oscillation startup and transistor switching. As a result, the phase noise is improved by a few dBs in comparison to the standard Colpitts architecture wherein a simple resistor or current source is placed in the emitters of the common base transistor pair.

The LC tank circuit that sets the oscillation frequency is placed in the collectors of the transistor pair. The frequency tuning range is divided in two sub-bands centered respectively on 9.75 GHz and 10.6 GHz. Band switching is implemented using

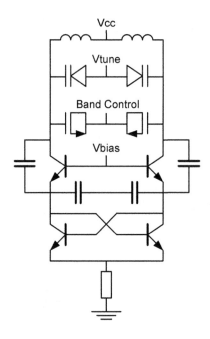

FIGURE 7.9

Simplified schematic of the RF VCO.

MOS varicaps, by taking advantage of their nonlinear capacitance-voltage characteristic that shows two constant-capacitance regions separated by an abrupt transition. Selection between low or high band is performed by setting the MOS varicap control voltage either at ground or supply voltage. Analog frequency tuning is performed using bipolar varicaps. In a rather unusual way, the varicaps are referenced to the VCO supply voltage to avoid the adverse effects of DC block capacitors and biasing resistors that are otherwise required when the varicaps are referenced to ground voltage. The PLL loop filter must also reference the VCO supply, which must therefore be available at a pin of the chip package. The tank circuit inductor is implemented in the top metal layer of the process. It has a symmetrical layout with a center tap that behaves as a virtual ground and where the voltage supply can therefore be applied for feeding current to the transistors, without disturbing the RF function. To minimize the risk of picking up unwanted radiated signals, the inductor has an eight-shaped layout which is proven to be less susceptible to magnetic radiations than single loop designs [14]. Thanks to the $3\,\mu$m-thick top metal, the inductor has a quality factor of about 25 at 10 GHz. As result, an overall quality factor of more than 10 is achieved over the tuning range that enables meeting phase noise requirements at current of a few mA.

The VCO is coupled to an amplitude control circuit (ACC) that regulates the oscillation level by controlling the oscillator core current. Amplitude regulation ensures constant output level in functions of frequency, temperature, and process variables, so that RF mixer and divider drives are maintained at optimum levels in all conditions. Furthermore, the ACC provides a means of starting the VCO at a higher current to achieve more reliable and faster oscillation startup.

The VCO buffers that complete the LO source function are implemented using cascoded differential amplifiers. Their role is to provide signal amplification and isolation from eventual input impedance modulations of the RF mixer or divider. Impedance modulations can be converted to unacceptable VCO frequency modulation if isolation is not sufficient. The mechanism by which impedance variations are converted to frequency variations is traditionally called load pulling. Modulations which are not synchronous with the LO frequency can be particularly annoying since they will show up as spurious signals in the VCO output spectrum. The RF divider, on the one hand, has potentially nonconstant input impedance, because it states changes several times in a reference cycle. The mixer, on the other hand, can show input impedance modulation at RF frequency, but only under large signal conditions. Furthermore, parasitic signals in the supply of the mixer or divider, such as residual tone at 22 kHz, can also cause unexpected impedance modulations. In our downconverter, this problem was prevented by using supply voltage regulators with more than 60 dB rejection at low frequency.

7.4.5 **Frequency divider**

The RF frequency divider has a modular architecture based on a cascade of asynchronous divide-by-2/3 cells [15], except for the front stage, which is a simple divider-by-2.

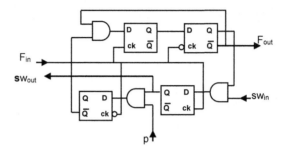

FIGURE 7.10

Circuit architecture of a divide-by-2/3 cell.

Compared to a divide-by-2 cell that requires only two D-latches, two more latches are necessary for implementing the control logic that determines division by 2 or 3 (Figure 7.10). The divider principle relies on the fact that each divide-by-2/3 cell is eventually asked to divide by 3 once in a division cycle, depending on its programming bit, p. Division-by-3 adds one more period of the cell input signal to the total division cycle. As the period added depends on the position of the cell in the chain, it follows that a chain of n divide-by-2/3 cells delivers an output signal with a period of:

$$T_{out} = \left(2^n + 2^{n-1} p_{n-1} + \cdots + 2p_1 + p_0\right) \times T_{in},$$

where T_{in} is the period of the input signal and $p_0 \cdots p_{n-1}$ are the programming bits of cells 1 to n, numbering from the input.

Division by 3 occurs when the swallow signal is high. Inversely to the RF divided signal, the swallow signal propagates backward, starting from the end of the divider chain. In order to ensure only one division by 3 in a complete division cycle, the swallow signal at each cell is designed to be high during only one period of the RF input signal of that cell. The swallow signal consists therefore of a pulse, which narrows by a factor of two when passed from one cell to the preceding. As the swallow signal pulse occurs once in a division cycle, it is generally taken as an output signal of the divider. The choice of where in the chain to take the swallow output depends on the desired balance between phase noise and pulse length. Closer to the input, the edges are of better quality, but the pulse is so narrow that the receiving circuit is imposed to be fast and hence rather current consuming too.

The divider cells are implemented in high-speed bipolar current mode logic (CML) for reasons explained above regarding the good ability of differential circuits to be integrated with other functions on the same die. To minimize the current consumption of the divider, the budget allocated to each cell was precisely adapted to its operating frequency. It follows that most current is consumed in the front cells. Furthermore, the current distribution between the different D-latches that compose a divider cell was carefully optimized for achieving the lowest consumption. More details on current optimization of divider cells can be found in [16].

7.4.6 Phase-frequency detector and charge-pump

The PFD is a combination of two edge detectors or D flip-flops plus an AND gate (Figure 7.11). A rising edge of either the reference or the divided VCO signal is detected, resulting in a logic 1 at the output of the corresponding flip-flop. When the other signal also has a zero crossing, both outputs will be logic 1, resulting in a reset. In the locked situation, the difference in time of the up and the down output will result in a net output current, proportional to the phase error between the reference XO and divided LO signal. When the phase error is 2π, the output current reaches its maximum average value, which is the DC value of the current sources.

This leads to the relation between input phase and output current:

$$K_{pd} = \frac{I_{cp}}{2\pi}. \tag{7.16}$$

In the locked situation, there will be small current peaks coming out of the up and down current sources. The uncertainty in the current level and the uncertainty in the timing of the edges will determine the phase noise at the input of the PFD.

Both PFD timing noise (jitter), δt_{rms}, and charge-pump current noise, $i_{noise}(f)$, can be translated back to the phase noise at the input of the PFD. For the jitter:

$$L(f_m) = 20 \cdot \log\left(\pi\sqrt{2} \cdot F_{REF} \cdot \delta t_{rms}\right). \tag{7.17}$$

And for the current:

$$L(f_m) = 20 \cdot \log\left(\pi\sqrt{2} \cdot \frac{i_{noise}(f)}{I_{cp}}\right). \tag{7.18}$$

Jitter is often the result of a slow zero crossing, typically at the input of the phase detector where the reference signal may be a sine wave with a low amplitude. Any noise at the input of the zero crossing detector will then directly transfer to jitter according to:

$$\delta t_{rms} = \frac{\delta v_{rms}}{dV/dt}. \tag{7.19}$$

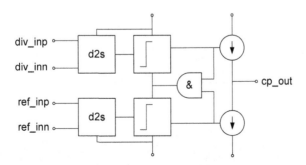

FIGURE 7.11

Block diagram of a phase-frequency detector and charge pump.

The noise of the charge-pump is that of the DC current source switched by the control signal. The noise power is therefore reduced in proportion to the duty cycle. As the duty cycle is commonly lower than 1% in a lock situation, the charge-pump noise power can be made very low. Then, if we assume that the original noise of the current source is white, the noise remains white after switching. It follows that the power spectral density of charge-pump noise current is also reduced in proportion to the duty cycle.

A logical choice for the phase-frequency detector is a CMOS implementation that has no static current except for the switching moments. As long as the slew rate within the logic cells is kept high, the influence of the PFD on phase noise can be minimal and will only be seen during the minimum "on" time of the charge-pump. A disadvantage of the single-ended implementation is the presence of high current peaks in the supply, which need to be filtered if reference spurs on the VCO are to be minimized.

The PFD is implemented using dynamic logic, which requires fewer gates than a static implementation. With proper dimensioning, fast operation with low current spikes can be achieved. The PFD is a single-ended circuit, but has differential inputs for both the reference and the divider input. Therefore the first stage of the PFD is a differential to single-ended (d2s) converter implemented as a comparator.

7.4.7 Charge-pump

The charge-pump does not have to deliver any current to the loop filter when in lock, but still the up and down switch will be on for a small amount of time. The implementation shown in Figure 7.12 uses a switched current source, which only needs a fixed bias voltage but no static DC current. The bias voltage can have some noise, since its effect will be canceled out at the output of the charge-pump. Current switching is performed by NMOS switches at the inputs and is done using the same devices in the up and down paths, leading to symmetrical behavior. The current pulses are mirrored using PMOS and NPN current mirrors and added to form the charge-pump

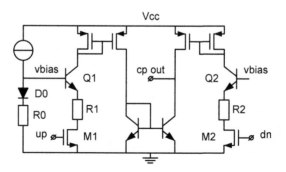

FIGURE 7.12

Circuit diagram of charge pump.

output. The role of the mirrors is also to have an output stage with low voltage drop current sources in order to maximize the output voltage range. The PMOS mirrors are rather slow but ensure nevertheless proper charge transfer, although the current waveform is quite distorted. In fact, their effect is equivalent to the addition of a high-frequency pole in the loop filter.

Because both the PFD and the charge-pump have very low static DC power consumption, the supply can be filtered with a relative high-value resistor in series, helping to suppress the current spikes resulting from the single-ended design style.

7.4.8 Crystal oscillator

The crystal oscillator sets the reference frequency and phase for the LO PLL. The term "reference" implies high frequency accuracy, stability, and low phase noise. To a large extent, these parameters are determined by the crystal resonator, which has very accurate and stable resonant frequency and a quality factor over 10,000, enabling the design of very low phase noise sources.

The crystal oscillator is implemented using a differential Colpitts core. The drawback of a differential architecture is that two pins are necessary for connection to the off-chip crystal. However, a differential architecture is less sensitive to common mode parasitic signals that can be eventually coupled to the XO via the substrate or bonding wires. The circuit diagram of the Colpitts XO is presented in Figure 7.13. Bipolar transistors were preferred to MOS for their lower $1/f$ noise. The common-collector

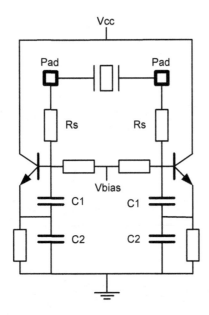

FIGURE 7.13

Simplified circuit diagram of the crystal oscillator.

configuration was chosen for its low voltage capability. In consequence, the crystal is placed differentially between the transistor bases. The oscillator operates on parallel resonance of the crystal with the output capacitance of the XO core. In a rather intuitive analysis, the oscillator operation can be understood considering a feedback loop amplifier model. The amplifier is the common-collector transistor which has voltage gain less than unity between its base input and emitter output. The feedback circuit consists of capacitors C_1, C_2 resonating with the crystal. Its voltage gain is $1 + C_2/C_1$, which can be made large enough for rendering the loop gain larger than unity, as is necessary for building an oscillator.

As the oscillator operates at parallel resonance, a natural approach for determining the startup conditions is to resort to a negative conductance model. The output conductance of the XO core is:

$$g_{xo} = -\frac{g_m}{2} \cdot \frac{\omega^2 C_1 C_2}{g_m^2 + \omega^2 (C_1 + C_2)^2}, \tag{7.20}$$

where g_m is the transconductance of one bipolar transistor. Oscillation startup requires:

$$g_{xo} + g_{xtal} < 0, \tag{7.21}$$

where g_{xtal} is the conductance presented to the core by the crystal at resonance. Thanks to the high-quality factor of the crystal, it can be shown that the conductance of the crystal is, when connected directly to the core:

$$g_{xtal} \approx r_m (c_0 + c_{core}) \omega_0^2, \tag{7.22}$$

where r_m is the motional resistance of the crystal and c_0, c_{core} are respectively the electrostatic capacitance of the crystal (that includes eventually any additional capacitance added on purpose in parallel to the crystal) and the output capacitance of the XO active core.

Fulfilling condition (7.21) puts constraints on the choice of C_1, C_2, and the core current. At 25 MHz, capacitor values in the range of 10–30 pF are typically required, which can still be easily integrated despite their relatively large value. In parallel, several 100 μA are necessary in the core for generating the required transconductance. The loss conductance g_{xtal} of the crystal also sets the load line for the transistor. It is generally desirable to have it under control for maintaining the oscillator level within desired limits. Depending on the crystal, Eq. (7.22) shows that the load line can vary in large proportions. A simple manner to desensitize oscillator operation from crystal parameters is to connect the crystal to the core via series resistors, typically in the order of 50 Ω. In this case, it can be shown that the conductance presented to the core is a bit larger but far less dependent on crystal parameters r_m, c_0. Contrary to what we could intuitively think, these series resistors have little impact on the quality factor at resonance. In addition, they contribute to increasing the robustness to electrostatic discharges that may eventually occur at connection pins.

7.5 **Experimental results**

Figure 7.14 shows a picture of the downconverter die where the location of the different functions is highlighted. The die size is about 1 mm². The area is about equally split between downconverter and PLL functions, respectively on right- and left-hand sides. Some noncritical functions such as regulators and ESD protections could be placed at the periphery of the die, between pads. Also embedded in the circuit is a I2C control bus that allows changing the default settings or measuring the DC operation point of some critical RF blocks. This was a very useful feature during the debugging phase for checking circuit operation and finding out the optimum settings. The I2C bus is not used in the application but is still employed in production testing. The right-band side of Figure 7.14 shows the downconverter in its 16-pin plastic package.

Figure 7.15 presents typical conversion gain and noise figure measured in low and high bands. The conversion gain is about 43 dB and the noise figure is below 7 dB

FIGURE 7.14

Photomicrograph of the downconverter die and view of the packaged IC.

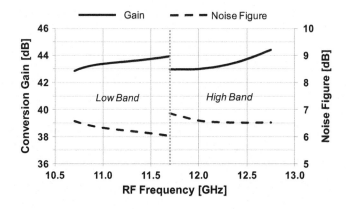

FIGURE 7.15

Measured downconverter gain and noise figure.

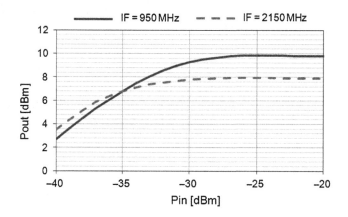

FIGURE 7.16

Compression characteristic of the downconverter.

in both bands. The positive gain slope implemented on purpose for compensating frequency-dependent attenuation in the IF cable is clearly visible. The downconverter has typical output IP3 of 16 dBm, 1 dB compression point of at least 6 dBm, and saturated output power over 8 dBm (Figure 7.16). The downconverter takes about half of the 52 mA consumed by the IC at room temperature. Comparison to earlier works in Table 7.4 reveals the step accomplished in reduction of current consumption.

Figure 7.17 shows typical PLL phase noise characteristic measured at IF output in low and high bands. The close-in noise is about −90 dBc on average and the integrated phase noise in a 10 kHz–13 MHz bandwidth is as low as 1° RMS in low band, and 1.3° RMS in high band, therefore significantly below the maximum specified limit of 1.7° RMS. Regarding the relatively low reference frequency (25 MHz) and the very low current consumption, this represents state-of-the-art phase noise performance. Figure 7.18 givess some details of the RF divider performance. Its sensitivity was measured in different biasing conditions using a standalone divider test circuit. The operating frequency range at −10 dBm input level is from 2 GHz to 22 GHz, giving a large enough margin with regard to the possible adverse effects of process spread and temperature variations on maximum operating frequency.

Table 7.4 Receivers Performance Comparison (Excluding PLL)

Reference	[2]	[17]	This Work
DC current	60 mA (est.)	75 mA	26 mA
Process	0.8-Bipolar	0.18-CMOS	0.25-BiCMOS
Gain	31±3 dB	50±2 dB	43±1 dB
NF	7.5±1.5 dB	3.5±0.6 dB	6.5±0.5 dB
Output IP3	16 dBm	17 dBm	16 dBm

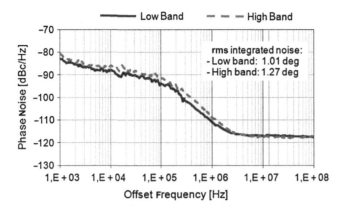

FIGURE 7.17

Measured PLL phase noise in low and high bands.

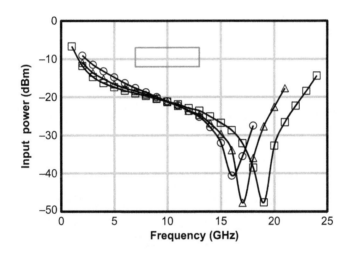

FIGURE 7.18

Measured RF divider sensitivity for various biasing conditions. Triangles: $Vcc=3.3\,V$ (nominal), squares: $Icc\,+30\%$, circles: $Vcc=3.1\,V$.

7.6 **LNB reference design**

The results shown in this section are taken from the application note describing the LNB reference design made by NXP [18]. The reference design is made with the four gain versions of the downconverter (TFF1014, TFF1015, TFF1017, and TFF1018).

7.6.1 **Design**

Figure 7.19 shows the circuit diagram of the LNB built around the downconverter IC. The other IC integrates the 5 V regulator and band/polarization control functions. The RF front-end is made of three discrete low noise transistors, followed by the hairpin band filter. Details of this filter printed on the circuit board are shown in Figure 7.20.

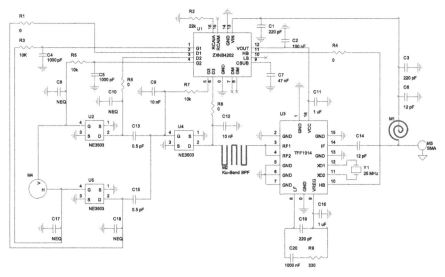

FIGURE 7.19

Schematic of LNB reference design.

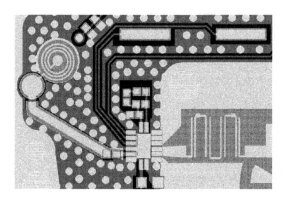

FIGURE 7.20

Details of the layout surrounding the downconverter. The hairpin filter is shown on the right. The via pattern and the trace location are essential for good performance.

FIGURE 7.21

Realized demonstration LNB. The feedhorn is replaced with a flange to accommodate measurements with an SMA to waveguide transition.

FIGURE 7.22

Details of printed circuit board without cover.

Figures 7.21 and 7.22 present respectively photographs of the demonstration LNB and of its printed circuit board.

7.6.2 Measurement results

Figure 7.23 shows the conversion gain and noise figure characteristics of the LNB using the 42 dB gain variant of the downconverter IC. The conversion gain of the LNB is about 61 dB and its noise figure around 1 dB.

One of the most relevant tests at LNB level consists of measuring the modulation error ratio (MER), defined as the ratio of the RMS powers of the received signal and the error vector. As modulation errors are caused by all kinds of imperfections of the receiver (such as noise, interferences, and distortion), the MER test gives a good indication of the overall quality of the LNB. Figure 7.24 shows the MER measured

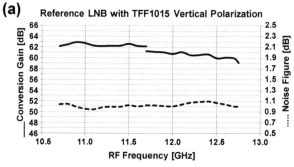

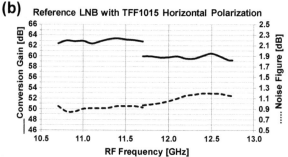

FIGURE 7.23

Typical conversion gain and noise figure characteristics of the LNB: (a) vertical polarization, (b) horizontal polarization.

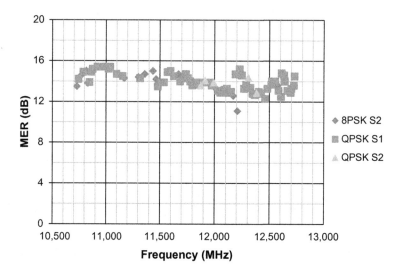

FIGURE 7.24

MER for N-PSK signal with NXP LNB #2 (TFF1015).

on real satellite signals using a DVB-S analyzer. The signal is received from Astra 19.2^E satellite using a 60 cm parabolic antenna. The weather conditions are almost blue sky and temperature of approximately 5 °C. The MER of 8PSK and QPSK meet Astra recommendations with level of about 14 dB for both modulations.

References

[1] C.S. Vaucher et al., Silicon-germanium ICs for satellite microwave front-ends, in *Proceedings of IEEE Bipolar/BiCMOS Circuits and Technology Meeting*, pp. 196–203, 2005.

[2] G. Girlando, S. A. Smerzi, T. Copani, and G. Palmisano, A monolithic 12-GHz heterodyne receiver for DVB-S applications in silicon bipolar technology, *IEEE Transactions on Microwave Theory and Techniques*, vol. 53, no. 3, pp. 952–959, 2005.

[3] Satellite link calculation. <http://www.satellite-calculations.com/Satellite/Downlink.htm>.

[4] Astra Universal Single LNB, Version 1.0, September 2007. <www.ses-astra.com>.

[5] P. Deixler et al., QUBiC4X: An fT/fmax = 130/140 GHz SiGe:C BiCMOS manufacturing technology with elite passives for emerging microwave applications, in *Proceeding of Bipolar/BiCMOS Circuits and Technology*, pp. 233–236, 2004.

[6] A. A. Abidi, General relations between IP2, IP3, and offsets in differential circuits and the effects of feedback, *IEEE Transactions on Microwave Theory and Techniques*, vol. 51, no. 5, May 2003.

[7] Stability analysis of linear periodical time-varying circuit using SpectreRF PSTB analysis, Application Note, Product Version 6.1, Cadence Design Systems, March 2006.

[8] E. van der Heijden et al., Low noise amplifier with integrated balun for 24 GHz car radar, in *IEEE Topical Meeting on SiRF*, pp. 78–81, 2008.

[9] C. Quemada, G. Bistué, and I. Adin, *Design Methodology for RF CMOS Phase Locked Loops*, section 2.2, p. 32. Artech House, Boston, USA, 2009.

[10] D. R. Stephens, *Phase-Locked Loops for Wireless Communications*, 2nd ed., section 13, p. 395. Kluwer Academic Publishers, New York, USA, 2002.

[11] E. van der Heijden et al., Colpitts VCOs for low-phase noise and low-power applications with transformer-coupled tank, in *Proceedings of the RFIC Symposium*, pp. 653–656, 2008.

[12] A. Fard and P. Andreani, An analysis of 1/f2 phase noise in bipolar colpitts oscillators, *IEEE Journal of Solid-State Circuits*, vol. 42, no. 2, February 2008.

[13] J.-P. Hong and S.-G. Lee, Low phase noise gm-boosted differential gate-to-source feedback colpitts CMOS VCO, *IEEE Journal of Solid-State Circuits*, vol. 44, no. 11, November 2009.

[14] O. Tesson, High quality monolithic 8-shaped inductors for silicon RF IC design, in *IEEE Topical Meeting on SiRF*, pp. 94–97, 2008.

[15] C. S. Vaucher et al., A family of low-power truly modular programmable dividers in standard 0.35-μm CMOS technology, *IEEE Journal of Solid-State Circuits*, vol. 35, no. 7, pp. 1039–1045, July 2000.

[16] C. S. Vaucher and M. Apostolidou, A low-power 20 GHz static frequency divider with programmable input sensitivity, in *IEEE RFIC Symposium*, pp. 235–23, 2002.

[17] Z. Deng, J. Chen, J. Tsai, and A. M. Niknejad, A CMOS ku-band single-conversion low-noise block front-end for satellite receivers, in *IEEE RFIC Symposium*, pp. 135–138, 2009.

[18] AN11444. Universal Single LNB with TFF101x FIMOD IC.

Bandpass ΔΣ ADCs for Wireless Receivers

Richard Schreier and Hajime Shibata

Analog Devices, Inc. Toronto, Canada

INTRODUCTION

Since ADCs were first tacked onto the end of a wireless receiver's signal chain the boundary between analog and digital processing has shifted steadily toward the antenna. As depicted in Figure 8.1, lowpass ADCs with bandwidths in the kHz range were initially used to digitize a single channel at baseband. Later, bandpass and sub-sampling ADCs did likewise at IF and, as the bandwidth and dynamic range of the ADCs improved, receivers increasingly relied upon digital signal processing to perform channel filtering. Nowadays, ADCs can digitize a 100 MHz swath of spectrum with sufficient dynamic range that multi-carrier operation with minimal filtering is possible. Analog-to-digital conversion is currently done at IF, but soon ADCs will be able to digitize GHz RF signals directly. When that day comes, the analog portion of a high-performance radio receiver could consist of only a band-select filter, an LNA, and an ADC. Going further, the LNA and even the filter may eventually be absorbed into the ADC so that fully programmable multi-band operation is possible.

In order for the dream of such a software-defined radio [1] to be realized economically, the ADC needs to be tunable. Tunability is required because power consumption is one of the most fundamental constraints on the ADC in a software radio, and digitizing signals that lie outside the band of interest squanders power. Bandpass ΔΣ ADCs [2–9] offer the ability to focus on the band of interest and therefore represent a promising technology for realizing this holy grail of ADCs.

At present, however, a multi-carrier receiver must mix the GHz RF signal down to a more manageable frequency. In a direct-conversion receiver a quadrature mixer translates the desired band to zero frequency, whereas in a superheterodyne receiver a (real or quadrature) mixer translates the desired band to a nonzero IF.

The specifications of the ADCs used in direct conversion appear to be less demanding than those of the ADC used in a superheterodyne receiver because both the signal frequency and bandwidth are lower. For example, to support a bandwidth of 100 MHz a direct-conversion receiver needs two ADCs to digitize the 0–50 MHz I and Q baseband signals, whereas a superheterodyne receiver needs a bandpass ADC supporting a 100 MHz bandwidth at an IF of perhaps 400 MHz. The lower frequency range used in direct conversion is one factor

Manganaro: Advances in Analog and RF IC Design. http://dx.doi.org/10.1016/B978-0-12-398326-8.00008-X

177

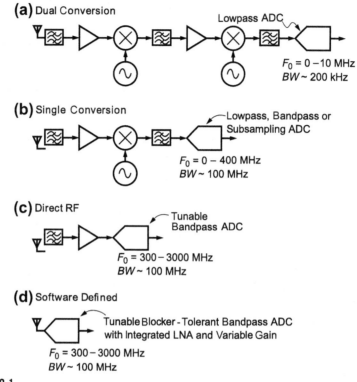

(a) Dual Conversion

Lowpass ADC

$F_0 = 0-10$ MHz
$BW \sim 200$ kHz

(b) Single Conversion

Lowpass, Bandpass or
Subsampling ADC

$F_0 = 0 - 400$ MHz
$BW \sim 100$ MHz

(c) Direct RF

Tunable
Bandpass ADC

$F_0 = 300 - 3000$ MHz
$BW \sim 100$ MHz

(d) Software Defined

Tunable Blocker - Tolerant Bandpass ADC
with Integrated LNA and Variable Gain

$F_0 = 300 - 3000$ MHz
$BW \sim 100$ MHz

FIGURE 8.1

Wireless receiver evolution.

that weighs in favor of that architecture. However, several other factors can tip the balance toward the superheterodyne approach, especially when a bandpass ADC is used.

First, note that the different ADC requirements are not necessarily significant. In theory, for a given noise density two ADCs of bandwidth $BW/2$ consume the same power as a single ADC of bandwidth BW. This rule holds in practice as long as BW is not so high as to come at a power premium. As the example given at the end of this chapter demonstrates, $BW = 100$ MHz does not stress the capabilities of 65 nm CMOS technology and an IF in the 200–400 MHz range can be supported without incurring an appreciable power penalty. Therefore the bandwidth and frequency-handling requirements of an IF-capable ADC may be inconsequential if a modern CMOS technology is used.

A second factor to consider is linearity. In a superheterodyne system with a sufficiently high IF, even-order distortion products fall out-of-band, whereas in a direct-conversion receiver even-order distortion terms can fall in-band. Particularly troublesome in a multi-carrier environment are the even-order distortion products that appear around DC since each carrier produces distortion terms near DC.

Table 8.1 Comparison of ADC Requirements for Direct Conversion and Superheterodyne Receivers

Criterion	Direct Conversion	Superheterodyne
Bandwidth	$BW/2$	BW
Maximum frequency	$\sim 1.5\,BW$	$\sim 5\,BW$
Even-order distortion	Sensitive	Insensitive
DC offset	Requires calibration and/or correction	Insensitive
$1/f$ noise	Problematic	Insensitive
Quadrature accuracy	Requires calibration and/or correction	Immune

For these reasons, the linearity requirements of the ADC are somewhat relaxed if the receiver architecture is superheterodyne rather than direct conversion.

In addition to even-order distortion, DC offset and $1/f$ noise also plague the low-frequency regime of a direct-conversion receiver. DC offset corrupts the channel at band-center and typically requires a combination of both calibration and digital correction to tame. DC offset can be sidestepped if frequency-planning is used to ensure that no active channel includes DC, but $1/f$ noise is still problematic. To overcome $1/f$ noise, either large devices or chopping must be used. In contrast, bandpass systems are insensitive to both DC offset and $1/f$ noise.

The final factor that weighs against direct conversion is quadrature accuracy. Imperfect quadrature in a direct-conversion receiver allows a signal with baseband frequency f to corrupt a weaker desired signal at $-f$. The required image attenuation is application dependent but can be in the order of 80 dB. To achieve this quality of quadrature, the gain and phase of the I and Q paths must match to within 0.002 dB and 0.01°, respectively, across the entire band. Adaptive digital correction is required since calibration alone cannot achieve the required accuracy, and the correction must be fast enough to track the variation of the correction terms with time.

Table 8.1 summarizes the above discussion. Although direct conversion minimizes the bandwidth and frequency range requirements of the ADCs, a superheterodyne receiver employing a bandpass ADC is safe from problems such as even-order distortion, DC offset, $1/f$ noise, and quadrature inaccuracy that afflict a direct-conversion receiver.

8.1 A bandpass $\Delta\Sigma$ primer

Figure 8.2 shows that a bandpass $\Delta\Sigma$ ADC consists of three main blocks, namely a loop filter, a coarse ADC, and a DAC. These blocks are connected in a feedback loop: the output of the loop filter is digitized by the coarse ADC whose output is in turn fed back via the DAC to the loop filter. By virtue of this arrangement, the output of the ADC provides good fidelity at frequencies where the gain of the loop filter is high. Since the linearity of the converter is limited by the linearity of the feedback DAC,

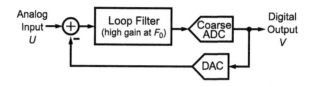

FIGURE 8.2

A simplified bandpass ΔΣ ADC.

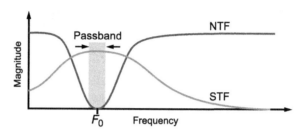

FIGURE 8.3

Bandpass noise and signal transfer functions.

rather than by the linearity of the internal ADC, and since high-speed current-mode DACs are more linear than plain high-speed ADCs, this configuration allows the linearity of high-speed ADCs to approach that of their DAC cousins.

Figure 8.3 shows a conceptual representation of the noise transfer function (NTF) and the signal transfer function (STF) of a bandpass ΔΣ ADC. The NTF is the closed-loop transfer function from the quantization error of the coarse ADC to the output of the ADC system. Since the in-band gain of the loop filter is high, the NTF is nearly zero in the band of interest and thus the quantization noise of the coarse ADC is highly attenuated in the passband. The STF is determined by the topology of the loop filter. The STF may have a weak bandpass shape as shown or it may have a less desirable shape with out-of-band peaks. In the context of a radio receiver subject to large out-of-band blockers, an STF with out-of-band peaks is clearly less attractive than one with out-of-band attenuation.

It is important to note that the *oversampling ratio (OSR)* of a bandpass ADC is

$$OSR = F_s/(2BW) \tag{8.1}$$

and thus *OSR* does not depend on the center frequency (F_0). For example, if the ADC uses a moderately high oversampling ratio of $OSR=32$ to digitize a bandwidth of $BW=100\,\text{MHz}$, the associated sampling rate is $F_S=6.4\,\text{GHz}$ even if F_0 is in the low-GHz range.

The remainder of this primer delves more deeply into ΔΣ ADCs by describing several of the key architectural choices (discrete time vs. continuous time, single bit vs. multi bit, etc.) and concludes with a discussion of the effects of clock jitter and phase noise.

8.1.1 Discrete-time vs. continuous-time ΔΣ

The mathematics of ΔΣ are best described using the language of discrete-time (DT) signal processing. DT ΔΣ ADCs implemented with switched-capacitor circuitry translate the DT difference equations directly into hardware. As long as the clock frequency is low enough that the circuits settle, the ADC obeys the difference equations and operates properly. In contrast, continuous-time (CT) ΔΣ ADCs are sensitive to the ratio of the clock period to various RC time-constants within the loopfilter and therefore require adjustable Rs or Cs in order to operate over a wide range of clock frequencies.

Nonetheless, realizing the loop filter of a ΔΣ ADC in CT form provides several advantages. Most importantly, the ADC system possesses inherent anti-aliasing, is power-efficient, operates at high speeds, is easy to drive, and readily supports variable gain. Furthermore, a CT bandpass ADC is able to exploit inductors in the realization of high-quality resonances with low power, low noise, and low distortion. Finally, as the design example at the end of the chapter shows, CT circuitry is able to take advantage of nm CMOS.

To understand the property of inherent anti-aliasing, consider the abstract single-loop CT system shown in Figure 8.4a. Since the feedback loop consists of a DAC driving a filter whose output is sampled, the feedback loop is a DT-input/DT-output linear time-invariant system and thus can be replaced by a DT equivalent as shown in Figure 8.4b. By observing that the closed-loop transfer function from the summation symbol to the ADC output is simply the NTF of the modulator, the STF is immediately found to be the product of the NTF and the open-loop transfer function of the loop filter driven from the analog input:

$$STF = L_{0c} \cdot NTF \tag{8.2}$$

This compact equation is formally correct but it hides the fact that STF is a mix of a continuous-time transfer function ($L_{0c}(s)$) and a discrete-time transfer function (NTF(z)). (The "c" subscript in L_{0c} indicates that the transfer function is continuous time.)

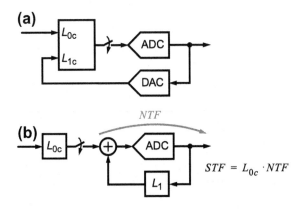

FIGURE 8.4

Derivation of the STF of a continuous-time ΔΣ ADC.

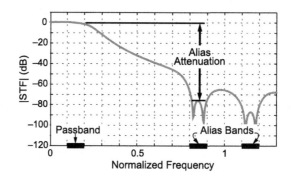

FIGURE 8.5

STF of a hypothetical CT bandpass ΔΣ ADC.

Strictly speaking, the arguments for these functions are incompatible but this technical problem can be resolved by only defining STF for physical frequencies, f:

$$\mathrm{STF}(f) = L_{0c}(j2\pi f) \cdot \mathrm{NTF}\left(e^{j2\pi f}\right) \tag{8.3}$$

Figure 8.5 plots the STF of a hypothetical CT ΔΣ ADC, showing how the periodic zeros of the NTF attenuate frequencies which alias to the passband.

8.1.2 Single bit vs. multi bit

One of the initial hallmarks of the ΔΣ field was the use of single-bit quantization. The reason for this early convention is that the two points of a single-bit DT DAC's input–output characteristic define a straight line and thus such a DAC is *inherently linear*. Multi-bit quantization provides much higher SQNR than single-bit quantization, but mismatch-induced nonlinearity was initially problematic. More recently, mismatch shaping is often used to shape the error caused by mismatch among the DAC elements in the DAC such that mismatch error is attenuated in the band of interest.

For the DACs used in CT ADCs, static element mismatch is only one of many DAC imperfections. Other impairments include static timing errors, inter-symbol interference, and a variety of other dynamic effects. Errors caused by static timing mismatch can be attenuated by mismatch shaping and switching errors can be attenuated by a modified form of mismatch shaping [10]. The other errors must be made sufficiently small by careful design, calibration, or digital correction.

8.1.3 Loop-filter architecture: feedback vs. feedforward

Figure 8.6 contrasts a loop filter that employs only feedback terms with one that employs only feedforward terms to control the zeros of L_1. Systems which use a

Feedback

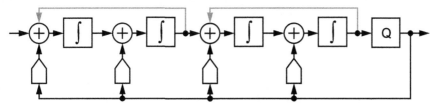

Feedforward

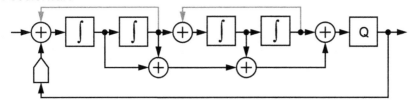

FIGURE 8.6

Feedback and feedforward loop filters.

combination of both feedback and feedforward exist, but for the sake of this discussion we consider only the canonical versions depicted in Figure 8.6.

The main reasons for preferring feedforward over feedback are power efficiency and simplicity. Since a feedforward system effectively replaces $n-1$ DACs in the feedback system with resistors, a feedforward system is simpler. Also, in a feedforward system the closed-loop transfer functions from the input to all integrators except the last have in-band nulls and thus the outputs of those integrators are devoid of in-band components. This property has beneficial consequences in terms of noise and distortion. Since nonlinearities acting on an integrator output do not operate on signal components, distortion of in-band signals is less problematic in a feedforward system than in a feedback system. To understand the noise implications, compare the first integrator in a feedforward system with that of a feedback system. The lower signal swings in the feedforward system allow the first integrator to have larger gain and thus the input-referred noise of subsequent stages is more attenuated in a feedforward system than a feedback system. As a result, for the same input-referred noise, less power needs to be allocated to the back-end in a feedforward system than in a feedback system.

The main reasons for preferring feedback over feedforward relate to the handling of out-of-band signals. As Figure 8.7 illustrates, the STF of a feedforward system invariably has out-of-band gain, whereas the STF of a feedback system can be designed to be flat in the vicinity of the passband with a gradual roll-off for frequencies outside the passband. The alias protection of a feedback system is also superior to that of a feedforward system.

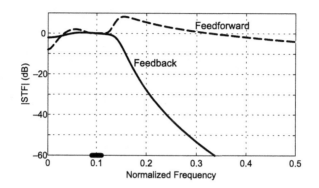

FIGURE 8.7

Example STFs for CT feedback and feedforward bandpass $\Delta\Sigma$ modulators.

8.1.4 Single-loop vs. multi-stage architectures

Figure 8.8 shows a simplified diagram of a multi-loop, or MASH, $\Delta\Sigma$ system. The idea behind this system is to use a second $\Delta\Sigma$ loop to digitize the quantization error of the first loop and to cancel this error with a digital cancelation filter. To see how this works, consider first the equations describing the two $\Delta\Sigma$ loops in Figure 8.8.

$$V_1 = \text{STF}_1 U + \text{NTF}_1 E_1, \tag{8.4}$$

$$V_2 = \text{STF}_2 W_1 + \text{NTF}_2 E_2. \tag{8.5}$$

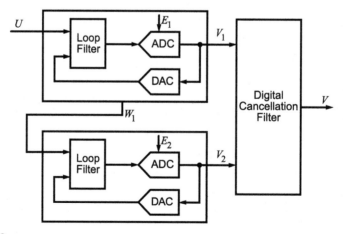

FIGURE 8.8

Two-loop $\Delta\Sigma$ modulator.

For the sake of simplicity, assume $\text{STF}_1 = \text{STF}_2 = 1$ and $W_1 = -E_1$ and consider the sum

$$V = V_1 + \text{NTF}_1 V_2. \qquad (8.6)$$

Substituting Eqs. (1.4) and (1.5) into Eq. (1.6) yields

$$V = U + \text{NTF}_1 \text{NTF}_2 E_2, \qquad (8.7)$$

which shows that the output of the system consists of the desired signal plus the quantization error of the second loop shaped by the product of the noise transfer functions of the two $\Delta\Sigma$ loops. Additional loops can be used to increase the order of the noise-shaping further.

The attractive feature of MASH is that a high-order NTF can be realized with the cascade of two or more low-order modulators. This approach allows more aggressive scaling with fewer stability concerns. The disadvantage of the MASH approach is that the effective NTF relies on cancelation. In order to achieve a high degree of cancelation, the loop filter typically needs to be implemented with switched-capacitor circuits, however recent work on adaptive filtering has allowed the use of CT loop filters [11]. Although it is not immediately obvious, it is worth noting that the alias protection in a MASH system is also tied to the overall NTF [12].

8.1.5 Jitter and phase noise

In a DT system, white sample-clock jitter of rms value σ_t results in a white error of rms value $e_{DT} = A\omega\sigma_t/\sqrt{2}$ when sampling a sine wave of amplitude A and angular frequency ω. In a CT system, jitter on the clock of the feedback DAC causes an rms error $e_{CT} = \Delta\upsilon_{rms}\,\sigma_t/T$ where $\Delta\upsilon_{rms}$ is the rms sample-to-sample change in the output of the modulator.

If the quantizer has a sufficient number of bits and the input amplitude and frequency are high, then $e_{CT} \approx e_{DT}$, i.e. the jitter susceptibility of a CT system is similar to that of a DT system. For example, if a bandpass CT system with center frequency F_0 has $\Delta\upsilon_{rms} = 2$, then $M > (\sqrt{2}/\pi)/(F_S/F_0)$ is sufficient to make its jitter susceptibility similar to that of a DT system. In lowpass modulators where the frequency is low, DT modulators generally fare better than CT systems in the face of clock jitter. It should be noted that in either case the in-band portion of the jitter-induced noise is $1/OSR$ of the total jitter noise power.

The above discussion applies to short-term variations in the clock period. Phase noise describes longer-term variations and manifests itself in the same way for both DT and CT systems. Figure 8.9 illustrates that the phase-noise skirts on the clock are replicated on the signal, scaled by F_{sig}/F_S. For example, if $F_{sig} = F_S/10$, the phase-noise skirts are shifted down by 20 dB.

In the remaining sections of this chapter, we present design details and measurements from a recent CT bandpass $\Delta\Sigma$ ADC.

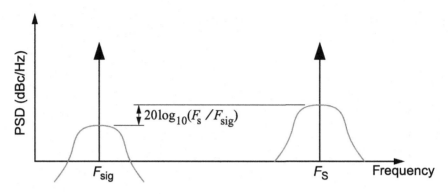

FIGURE 8.9

Phase noise on the clock shows up as phase noise on the signal, scaled by the ratio of clock and signal frequencies.

8.2 Example bandpass ΔΣ ADC

The architecture of the tunable bandpass ΔΣ ADC that will be discussed in detail is shown in Figure 8.10 [2]. The ADC consists of a sixth-order continuous-time tunable lowpass/bandpass loop filter, a 17-level quantizer and seven feedback DACs clocked at 2–4 GHz. With a clock frequency of 4 GHz, the center frequency of the ADC can be tuned from DC to 1 GHz. The bandwidth is also adjustable, with a useful range of 30–150 MHz. The continuous-time feedback-style loop filter allows the STF to have a large amount of alias attenuation and to be free of out-of-band peaks. As described earlier, these properties are advantageous in the context of wireless communication applications in which large blocker signals are present. To support the bandwidth (>100 MHz) and dynamic range (>70 dB) requirements of emerging radio communication systems, a sixth-order loop filter and a 17-level quantizer were selected.

The input signal is supplied either from the on-chip LNA or attenuator (ATT). Each block supports a 50 Ω input impedance and when powered down presents an

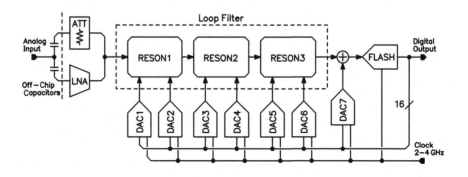

FIGURE 8.10

ADC architecture.

open circuit at its input and output ports. The LNA covers the full-scale (FS) range from −18 dBm to −6 dBm while the attenuator covers the range from −6 dBm to +18 dBm, for a total gain range of 36 dB.

The output of the ADC is a 4 b 2–4 GHz digital signal. This signal is processed by digital decimation filters and digital down-conversion mixers to yield a baseband signal devoid of mixer artifacts.

As described earlier, a feedback-style loop filter has signal content at the integrator outputs, and therefore needs low-distortion integrators in order to achieve low distortion for the ADC as a whole. In this ADC the op amps need an in-band gain in excess of 50 dB in order to achieve distortion below −80 dBc. Section 8.2.3 describes how such a high amplifier gain is achieved at 1 GHz signal frequencies.

In multi-bit ΔΣ ADCs, a mismatch shaper is typically inserted in the digital feedback path between the quantizer and the feedback DACs. Although this approach would be effective in minimizing the in-band degradation caused by mismatch in the DAC elements, it adds more delay to the feedback path than is tolerable at a 4 GHz clock rate and is also impractical considering the tunable nature of the ADC. Section 8.2.5 describes the DAC circuit used to achieve the required DAC linearity.

8.2.1 Input LNA and attenuator

Figures 8.11 and 8.12 show simplified LNA and attenuator schematics. The LNA utilizes a noise-canceling architecture [13] modified to produce a current output and to be compatible with the attenuator. The LNA consists of a programmable number of common-gate (CG) slices in parallel with a programmable number of common-source (CS) slices. The CG slices match the LNA to the source impedance, R_S, and yield a transconductance of $1/R_S$. Additional transconductance, up to a total of $4/R_S$, is achieved with the CS stages. By turning off the CG and cascode transistors, both the LNA output and its input become high impedance, thereby allowing the LNA to be placed in parallel with the resistive attenuator. The circuit was designed to

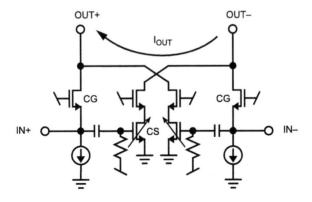

FIGURE 8.11

Simplified LNA schematic.

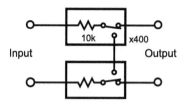

FIGURE 8.12

Simplified attenuator schematic.

accommodate source resistances from 50 to 200 Ω, although when Z_{in} is high the high-frequency matching is degraded due to the capacitive overhead associated with the high degree of programmability.

8.2.2 Loop filter

The loop filter of a ΔΣ ADC provides the noise-shaping. In tunable bandpass ΔΣ ADCs, the resonance frequencies of the loop filter are tuned to frequencies within the band of interest to attenuate quantization noise in that band. Figure 8.13 shows an example loop filter responses for center frequencies ranging from DC to 1 GHz.

This widely tunable loop filter was implemented with a hybrid LC/active-RC structure. A single-ended representation of the loop-filter is shown in Figure 8.14. Three operating modes, namely the active-RC, LC1, and LC2 modes, cover the [0, 200], [200, 500], and [500, 1000]-MHz center-frequency ranges, respectively.

In the active-RC mode, the selection cascode transistor S0, amplifiers A1, A2, A3a, and A4a, as well as DACs DAC2, DAC3a, and DAC4a are activated as shown

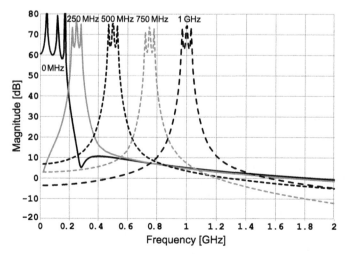

FIGURE 8.13

Target loop-filter frequency responses for $F_0 = 0, 250, 500, 750, 1000$ MHz.

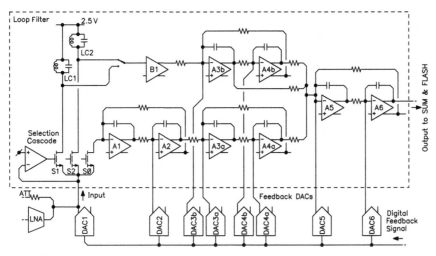

FIGURE 8.14

Structure of the loop filter.

Table 8.2 Loop-Filter Sub-Blocks Activated in Each Operating Mode

Mode	Selection Cascode	RESON 1	RESON 2
Active-RC	S0	A1, A2, DAC1, DAC2	A3a, A4a, DAC3a, DAC4a
LC1	S1	LC1, B1, DAC1	A3b, A4b, DAC3b, DAC4b
LC2	S2	LC2, B1, DAC1	

in Table 8.2. In the LC1 and LC2 modes, the selection cascode associated with either the LC1 or LC2 tank, buffer B1, amplifiers A3b and A4b, plus DACs DAC3b and DAC4b are activated. Amplifiers A5 and A6, and DACs DAC1, DAC5, and DAC6 are used in both active-RC and LC modes.

The LC resonators employ external inductors, having values of 20 nH for LC1 and 5 nH for LC2, to exploit the superior linearity and power efficiency of an LC tank containing high-Q off-chip inductors. Tuning of the tank's resonance frequency is accomplished with a 12-bit programmable on-chip capacitor bank supporting capacitance ranges of 5–30 pF for LC1 and 4–20 pF for LC2.

The first resonator in the active-RC mode is a traditional active-RC-based resonator using two feed-forward op amps because an LC-based resonator does not provide the lowpass resonance needed to support the lowpass ΔΣ ADC configuration. A 2.5 V/1.0 V dual-supply op amp is employed for the A1 and A2 amplifiers in the first resonator to minimize the noise from the inter-stage resistors by maximizing the output voltage swing of the A1 and A2 amplifiers.

When the ADC is in bandpass mode, it would be beneficial to use LC-based resonators in the back-end, but this approach requires more pins to accommodate external inductors while internal inductors do not provide high enough Q at the required

inductance range. Since LC resonators are ruled out by the proceeding considerations, the second and third resonators are traditional active-RC-based resonators in
both the active-RC and LC modes. A feed-forward op amp using a 1 V supply voltage
is employed to maximize bandwidth and minimize the power consumption of the
A3–A6 amplifiers in these back-end resonators. The reduced output swing is tolerable because the noise requirements for these stages are relaxed.

The resistor and capacitor component values in the loop filter, the feedback DAC
currents, and the flash full-scale are digitally programmable with 8- to 12-bit resolution. Typical component ranges are $100\,\Omega$–$20\,k\Omega$ for resistors, 0.8–4.0 pF for the
capacitors, and 0–512 μA for the DAC I_{LSB} currents. These parameter settings are
calculated off-line using a combination of device characterization and theoretical
calculations based on the desired NTF.

8.2.3 Amplifier

Two op amp designs are used in the loop filter: a 2.5 V/1 V dual-supply design is used
in A1 and A2 and a 1 V single-supply design is used in A3–A6. Both designs are
multi-stage multi-path feed-forward op amps [14–18].

The basic concept of a multi-stage multi-path feed-forward op amp is to combine the
high bandwidth of a single-stage amplifier with the high gain of a multi-stage amplifier by
summing their outputs. Figure 8.15a shows a basic third-order feed-forward op amp. The
first-order path in this amplifier consists of a single stage (g_{m1}) while the third-order path
consists of three stages (g_{m3A}, g_{m3B}, and g_{m3C}). A second-order path (g_{m2A} and g_{m2B})
provides a controlled transition from third- to first-order behavior. The overall gain of
the amplifier is the envelope of the path gains as shown in Figure 8.15b because the path
with the highest gain dominates the response of the amplifier at any given frequencies.

To make the amplifier stable in a feedback loop, the gain must be gradually
reduced from a high-order response to a first-order response around the unity gain

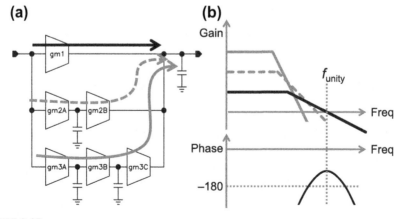

FIGURE 8.15

Conceptual diagrams of a multi-path multi-stage feed-forward op amp.

frequency so that the phase response approaches 90° as shown in Figure 8.15b. Inter-stage capacitors which control the bandwidths of the second- and higher-order paths ensure that the amplifier is closed-loop stable.

Since a multi-stage feed-forward op amp is a conditionally stable system, the amplifiers could become unstable if the signal swing is large enough to cause the effective gain of the internal stages to be reduced. This behavior is problematic if the amplifiers are used in an ordinary filter, but is less problematic in a $\Delta\Sigma$ ADC. The reason is that a $\Delta\Sigma$ ADC, which is also a conditionally stable system, also becomes unstable in such a large-state scenario and must be reset. Resetting the amplifiers whenever the ADC is reset is sufficient to solve the problem of conditional stability in the amplifiers.

The amplifiers in a $\Delta\Sigma$ ADC must satisfy two gain requirements, one in the vicinity of the signal band and one in the vicinity of $F_S/2$. In the signal band, the amplifier gain must be high enough to meet the distortion targets and to ensure sufficient coefficient accuracy. A typical requirement is 40–50 dB of loop gain at the signal frequency for feedback-style $\Delta\Sigma$ ADCs or as low as 20 dB of loop gain for feedforward-style $\Delta\Sigma$ ADCs. At $F_S/2$ a moderate loop gain of 10–20 dB is needed to process the injected current from the feedback DACs. In bandpass feedback-style $\Delta\Sigma$ systems, the first requirement is usually more stringent because the in-band frequency F_0 is typically within a few octaves of $F_S/2$.

In this system, −80-dBc distortion and 1% coefficient variation require 50 dB of gain at 200 MHz for A1–2 and 40 dB of gain at 1 GHz for A3–6. These requirements translate into gain bandwidth (GBW) products of 63 GHz and 100 GHz. Since these GBW requirements are almost the same as the maximum f_T of the 65 nm CMOS devices used in this design, a traditional op amp design having a first-order roll-off would be unable to meet the GBW requirements. However, in a feed-forward op amp, the low-frequency gain and the unity-gain frequency are decoupled by the high-order roll-off. Since a GBW product of only 6 GHz or so is needed at $F_S/2$ and since the low-frequency GBW can be increased by adding higher-order paths, a feed-forward op amp is a good fit for the requirements of a high-speed feedback-style bandpass $\Delta\Sigma$ ADC.

A simple extension of the third-order design shown in Figure 8.15a into a fifth-order design yields the amplifier design shown in Figure 8.16a. This design is power-inefficient because five independent output stages consuming similar amounts of power are connected in parallel. To make the amplifier more power efficient, stage sharing and power scaling can be applied as illustrated in Figure 8.16b. In this circuit, four output stages are merged into g_{m2}, two output driver stages are combined into g_{m6}, and three input stages become g_{m4}. In a similar manner to scaling in CMOS inverter chains, power scaling allocates less power to stages that are far from the power-hungry output stage. In this design, the output stage consumes 38 mW whereas internal stages consume as little as 2 mW; 14 mW is allocated to the g_{m4} input stage for noise reasons. The total power consumption of the A1 amplifier is 100 mW while the sum of the input and the output stage power consumption is 49 mW, which is essentially the minimum power consumption needed to achieve the required input-referred noise and load-driving capability. Therefore the power penalty associated with the gain requirements relative to the drive and noise requirements is a factor of two for this fifth-order amplifier.

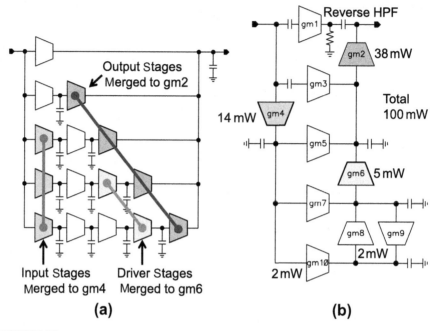

FIGURE 8.16

Fifth-order multi-stage multi-path feed-forward op amps: (a) primitive fifth-order design, (b) A1 op amp design.

To minimize power consumption, the 2.5 V supply voltage is only used in the main output stage, g_{m2}, and the main input stage, g_{m4}. The output stage, g_{m2}, shown in Figure 8.17 is a 2.5 V design to maximize the output voltage swing. This stage consists of complementary 1 V NMOS/PMOS pseudo-differential pairs to maximize g_m

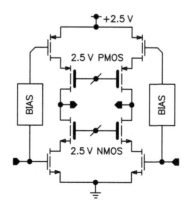

FIGURE 8.17

Output stage of the A1 op amp.

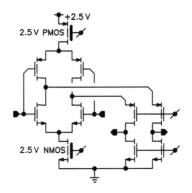

FIGURE 8.18

Input stage of the A1 op amp.

and 2.5 V cascode devices (denoted with a thick gate) to protect the 1 V devices. The input stage, g_{m4}, shown in Figure 8.18 is a folded-and-telescopic g_m stage that has PMOS and NMOS differential pairs connected in parallel to maximize the g_m/I_{bias} ratio. The common-mode voltage is reduced from 1.25 V at the input to 0.7 V at the output to avoid over-voltage stress in the subsequent 1 V stages. Other internal stages employ differential pairs or pseudo-differential pairs using 1 V MOSFETs to maximize bandwidth and minimize power consumption.

In addition, the first-order and second-order bypassing stages, g_{m1} and g_{m3}, are AC-coupled in order to interface these 1 V circuits to the 1.25 V I/O common-mode voltage. A reverse-connected high-pass filter at g_{m1}'s output protects g_{m1} from the large-voltage swing present at the amplifier output. Lastly, stages g_{m8} and g_{m9} produce a gain resonance at 300 MHz to enhance the high-frequency gain of the amplifier.

The design used for the A3–A6 amplifiers, which handle signal frequencies up to 1 GHz, is a pure 1 V supply seventh-order multi-stage feedforward amplifier employing a structure similar to that of the fifth-order design.

The loop gains of both designs are shown in Figure 8.19. The plot for A1 confirms that five g_m stages connected in series provides 60 dB of gain at 250 MHz for a 250 GHz effective GBW with a unity-gain frequency of 6 GHz. The A3 design achieves 40 dB at 1.5 GHz for a 150 GHz effective GBW, and a 15 GHz unity-gain frequency. The plots also illustrate the loop gain has a fifth- or seventh-order response in the low-frequency region and reverts back to a first-order response as the unity gain frequency is approached. The loops are conditionally stable having phase margins of 75° and 65°, respectively.

8.2.4 Flash ADC

The clock frequency of a ΔΣ ADC is typically limited by the digital feedback path, specifically by the regeneration time-constant (τ) of the comparator in the internal flash ADC and the propagation delay from the flash output to the DAC input. As shown

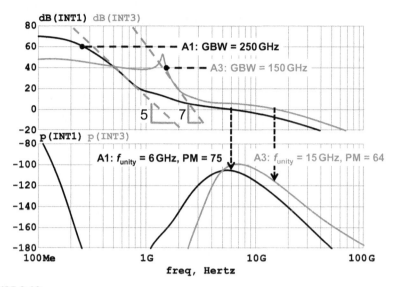

FIGURE 8.19

Simulated A1 and A3 op amp loop-gain.

in Figure 8.20, regeneration, propagation, and sufficient setup time for the feedback DACs must all fit within one 250 ps clock cycle. Increasing the current densities and widths of the cross-coupled devices used in the comparator core reduces τ and reduces the number of inverters in the DAC driving chain, but increases power consumption.

In this design, we employ a triode PMOS as the load for cross-coupled NMOS devices in the regeneration core of the comparator (Figure 8.21). A differential current is provided as the input signal to the comparator core. The use of triode PMOS devices as the load makes V_{GS} of the NMOS devices higher than $V_{DD}/2$ when regeneration starts, thereby minimizing τ. However, with only a triode PMOS device, one of the cross-coupled NMOS devices conducts feed-through current even after the decision is fully made because both the PMOS load device and the bottom NMOS device conduct in one of the branches.

This feed-through current can be minimized by introducing a self-cut-off latch connected at the gate of the PMOS load devices. In the tracking phase where the clock signal, CK1, is low, the PMOS gate terminals, Yp and Yn, are set to low by the switches, thereby turning on the PMOS devices in the tracking phase. When CK1 goes high, the comparator core goes into the regeneration phase, where the regeneration

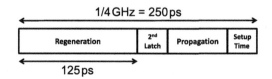

FIGURE 8.20

Timing allocation in the digital feedback path.

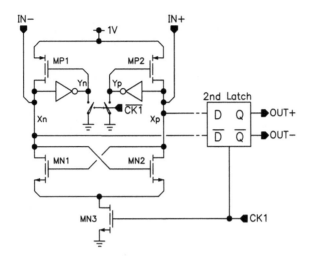

FIGURE 8.21

Simplified schematic of the self-cut-off comparator.

nodes, X_p and X_n, are pulled from V_{DD} down to $V_{DD}/2$ and start to diverge. During this interval, the inputs of the inverters are both high, and the PMOS devices are on. As the regeneration proceeds, the X_p and X_n voltages diverge and approach V_{DD} or V_{SS}. In the case X_p goes up to V_{DD} and X_n goes down to V_{SS}, MN2 on the right branch will be turned off because V_{GS} of MN2 goes to zero. On the left branch, the low voltage X_n causes the Y_n voltage to change from V_{SS} to V_{DD} through the inverter. This action turns off MP1 and thus shuts off the left-branch current. Therefore no feed-through currents flow in the comparator core when the regeneration is fully completed. With these techniques the comparator nominally achieves $\tau = 6\,\text{ps}$ with a power consumption of $3\,\text{mW}$ per comparator at a clock frequency of $4\,\text{GHz}$.

8.2.5 Feedback DACs

Each comparator in the flash ADC is connected though a second latch and an inverter chain to a DAC element as shown in Figure 8.22. Two clock signals, CK1 and CK2, depicted in Figure 8.23, implement the [1,2] DAC timing. First, CK1's rising edge

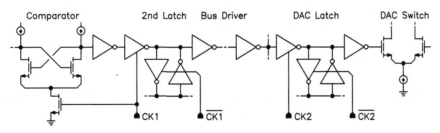

FIGURE 8.22

Flash-to-DAC path.

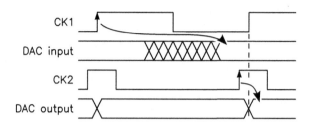

FIGURE 8.23

Flash and DAC clock timing.

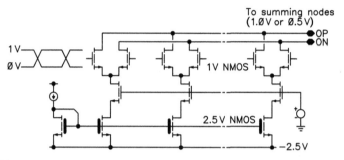

FIGURE 8.24

DAC switch and current source simplified schematic.

triggers the sampling action in the comparators. The comparator output then propagates to the DAC input after the regeneration and the propagation delays. After the data bits settle, CK2 rises, the DAC latches become transparent, and the outputs of the DACs are updated. CK2's or CK1's rising edge is adjusted to align the sampling and DAC-update instants.

The drive strength of the second latch is increased by the inverter chain connected to the latches in DACs 1–7. The DAC latches and the DAC drivers use full-swing 0–1 V CMOS levels for simplicity and high speed.

The DAC structure shown in Figure 8.24 consists of a differential switch pair on the top, cascode transistor in the middle and a current-source device at the bottom. A large voltage of 3.0–3.5 V is present from the DAC output nodes to the −2.5 V supply voltage. This arrangement allows a high effective voltage (V_{eff}) for each device. For the switch pair a high V_{eff} results in good dynamic performance and low output parasitic capacitance, while for the 2.5 V NMOS current sources a large V_{eff} results in low noise and good matching.

8.3 Experimental results

The ADC is fabricated in a mixed-signal 65 nm CMOS process having 1.0 V core and 2.5 V I/O devices. A die photo of the fabricated chip is shown in Figure 8.25. The area of the modulator including the loop filter, flash, DACs, and bias is 5.5 mm². When

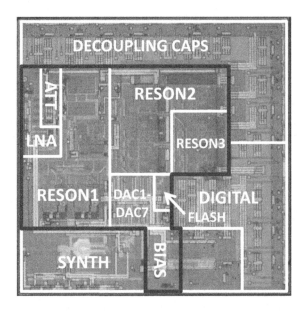

FIGURE 8.25

Die photo. The ADC area is enclosed by the black lines.

clocked at 4 GHz, the ADC consumes 550 mW in the two LC modes and 750 mW in the active-RC mode.

8.3.1 NTF and STF

Figures 8.26, 8.27, and 8.28 show measured output spectra and STFs for $F_0 = 1$ GHz, 450 MHz, and 0 MHz, demonstrating wide center-frequency tunability from 0 to $F_S/4$ with no STF peaking. The STF roll-off at low frequencies in the $F_0 = 0$ MHz case is due to an off-chip balun and off-chip AC-coupling capacitors. For the preceding values of F_0, dynamic range and peak SNR are 67/79/73 dB and 63/72/71 dB, respectively.

The noise spectral density (NSD) of the ADC depends on the BW setting because the back-end noise degrades the NSD at the inband edges in wide-band configurations. The ADC achieves a mean noise density of -159/-156/-152 dBFS/Hz for BW of 50/100/150 MHz at $F_0 = 400$ MHz as shown in Figure 8.29. Beyond $BW = 150$ MHz, the increase in NSD at the band edge is too great to be of practical value.

By setting the BW to the impractical value of 300 MHz, however, the notches of the NTF become clearly visible, which allows the resonator, Q, to be determined. Figure 8.30 shows that the Q of the active-RC resonators is high enough that the notch width is less than 25 MHz. Thus the resonator Q is high enough to prevent degraded noise shaping as long as the BW is more than 25 MHz. Another factor which limits the achievable noise shaping is the frequency accuracy of the NTF zeros. This design

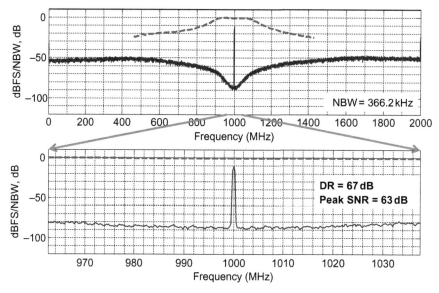

FIGURE 8.26

Measured single-tone output spectrum at $F_0 = 1\,\text{GHz}$ and $BW = 75\,\text{MHz}$. Dashed and solid lines represent STF and output spectra.

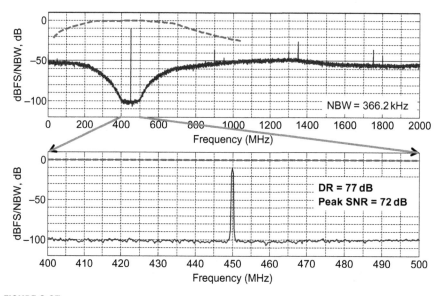

FIGURE 8.27

Measured single-tone output spectrum at $F_0 = 450\,\text{MHz}$ and $BW = 100\,\text{MHz}$. Dashed and solid lines represent STF and output spectra.

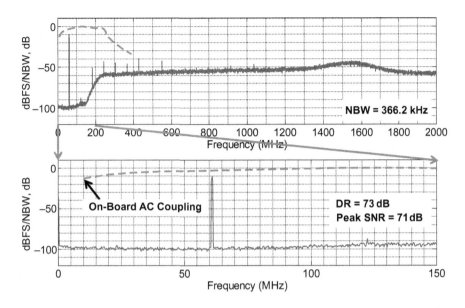

FIGURE 8.28

Measured single-tone output spectrum at $F_0 = 0$ MHz and $BW = 150$ MHz. Dashed and solid lines represent STF and output spectra.

uses a combination of N-poly (positive temperature coefficient) and P-poly (negative temperature coefficient) resistors to reduce the temperature-dependence of the resistors. Temperature sweeps confirm that the NTF zeros shift by less than 1% over the -40–$125\,°C$ range.

In theory, the alias attenuation afforded by this feedback ADC exceeds 130 dB. In practice, feed-through in the integrators and other leakage paths reduce the

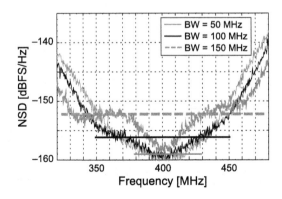

FIGURE 8.29

NSD at various BW; $F_0 = 400$ MHz, $F_S = 4$ GHz.

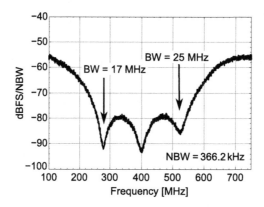

FIGURE 8.30

Measured output spectrum at $F_0 = 400$ MHz and $BW = 300$ MHz showing resonator Qs.

high-frequency roll-off of the loop filter. Measurements indicate that the alias attenuation is roughly 80 dB.

8.3.2 Single-tone and two-tone performance

Figure 8.31 and Figure 8.32 show single-tone and two-tone in-band spectra for $BW = 75$ MHz around $F_0 = 350$ MHz. Spurs are below -95 dBFS in the single-tone

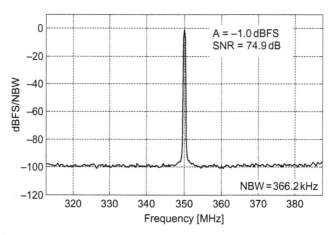

FIGURE 8.31

Measured single-tone spectrum at $F_0 = 350$ MHz and $BW = 75$ MHz.

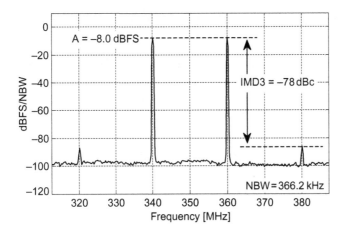

FIGURE 8.32

Measured two-tone spectrum at $F_0 = 350$ MHz and $BW = 75$ MHz.

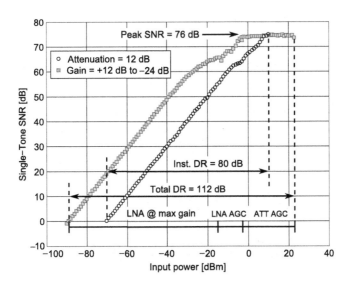

FIGURE 8.33

Measured SNR vs. input power with 350 MHz input signal at $F_0 = 350$ MHz, $BW = 75$ MHz.

case, while IMD3 < −78 dBc with two −8 dBFS tones. A single-tone amplitude sweep measured at the same settings is depicted in Figure 8.33. In this configuration, the ADC achieves a peak SNR of 76 dB and an instantaneous dynamic range of 80 dB. Including the 36 dB gain range of the LNA and attenuator extends the dynamic range to over 110 dB.

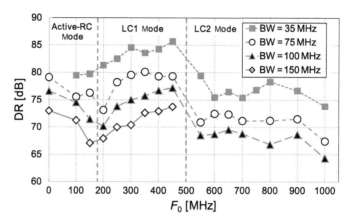

FIGURE 8.34

Measured instantaneous dynamic range vs. center frequency, F_0.

8.3.3 *DR* vs. F_0 and *BW*

The instantaneous dynamic range at a fixed full-scale of +6 dBm for several band-width and center-frequency settings is plotted in Figure 8.34. The increasing dynamic range with increasing F_0 seen in the LC1 mode is due to the fact that the gain of the first resonator increases and thereby suppresses the noise from the back-end components. At $F_0 = 450$ MHz, 86 dB of instantaneous dynamic range is achieved with a narrow-band ($BW = 35$ MHz) configuration, while $DR = 74$ dB is achieved in a wide-band ($BW = 150$ MHz) configuration. For $BW = 75$ MHz, 80 dB of dynamic range is achieved.

8.3.4 Comparison with other ΔΣ ADCs

The ADC performance for selected settings is summarized in Table 8.3 along with other recently reported bandpass and lowpass ΔΣ ADCs. The figure of merit (FOM) used in the table is a power-efficiency metric defined as

$$\text{FOM} = DR + 10 \log_{10} \frac{BW}{P} \tag{8.8}$$

where DR is instantaneous dynamic range in decibels, BW is bandwidth of the ADC in Hertz, and P is power consumption of the ADC in Watts, excluding back-end digital signal processing [19]. As shown in the table, this ADC has the widest bandwidth among bandpass and lowpass ΔΣ ADCs reported with $DR > 60$ dB.

Figure 8.35 is a FOM vs. bandwidth plot including all ADC data points reported over the past 16 years at the International Solid-State Circuits Conference (ISSCC) [20]. Open circles represent non-ΔΣ ADCs, gray circles represent ΔΣ ADCs, and squares mark ΔΣ ADCs listed in the references. The data points of this ADC

Table 8.3 Performance Summary With Recently Reported ΔΣ ADCs. BP/LP and FB/FF Represent Bandpass/Lowpass and Feedforward/Feedback Architecture

	This work			Yamamoto JSSC 2008 [9]	Lu JSSC 2010 [7]	Harrison ISSCC 2012 [6]	Chae ISSCC 2012 [8]	Ryckaert JSSC 2009 [3]	Chalvatzis JSSC 2009 [4]	Luh VLSI 2006 [5]	Shettigar ISSCC 2012 [21]	Bolatkale ISSCC 2011 [22]
	Active-RC	LC1	LC2									
Architecture	FB LP	FB BP		FF LP/BP	FP BP	FP BP	FP BP	FP BP	FP BP	FF BP	FB LP	FF LP
Inst. DR(dB)	79 77 73	80 77 74	71 69 74	93 76	70	70	60	45	54	62 48	80	70
BW (MHz)	75 100 150	75 100 150	75 100 35	0.31	10	20	24	60	120	60 180	36	125
Power (mW)	750	550		115	160	20	12	40	16,000	7,700	15	256
FOM (dB)	159 158 156	161 160 158	152 151 151	157 140	148	160	153	137	133	131 122	174	157
F_0 (MHz)	0	450	550 1,000	0 12.6	200	700–800	200	2,400	2,000	1,400	0	0
Process	65nm CMOS			180nm CMOS	180nm CMOS	40nm CMOS	65nm CMOS	90nm CMOS	130nm SiGe	InP HBT	90nm CMOS	45nm CMOS
V_{DD} (V)	1.0/±2.5			1.8	1.8		1.25	1	2.5	±5/−3.3	1.2	1.1/1.8
F_s (GHz)	4.0			0.04	0.8	3.2	0.8	3.0	40	4.0	3.6	4.0

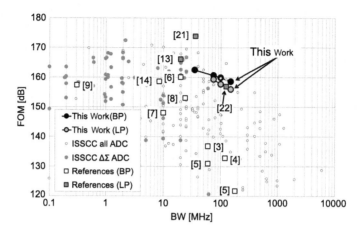

FIGURE 8.35

FOM vs. *BW* in ISSCC 1997–2012 [20].

are large black dots connected with solid lines to indicate the variable bandwidth feature. The upper group of points corresponds to the LC1 mode and the lower group corresponds to the active-RC mode. As this plot shows, this ADC has a power efficiency which is competitive with other ADCs in the 100 MHz bandwidth region regardless of architecture. In addition to having competitive power efficiency, this ADC also possesses features such as center-frequency tunability, bandwidth reconfigurability, built-in anti-aliasing, a bandpass STF with no out-of-band peaking, and a variable full-scale, which are not generally supported in other types of ADCs.

8.4 Looking forward

The trend in high-performance receivers is to do analog-to-digital conversion as early as possible in the signal chain. In the extreme case the ADC is attached directly to the antenna. This chapter showed that continuous-time bandpass ΔΣ is a promising architecture for this ultimate ADC and presented design details and measurement results from a recent highly-programmable implementation. In the authors' opinion, the day when ADCs digitize RF signals up to a few GHz should be just around the corner.

References

[1] J. Mitola, The software radio architecture, *IEEE Communications Magazine*, vol. 33, no. 5, pp. 26–38, May 1995.

[2] H. Shibata et al., A DC to 1 GHz tunable ΔΣ ADC achieving DR = 74 dB, BW = 150 MHz and f_0 = 450 MHz using P = 550 mW, in *International Solid-State Circuits Conference Digest of Technical Papers*, pp. 150–152, February 2012.

[3] J. Ryckaert et al., A 2.4 GHz low-power sixth-order RF bandpass $\Delta\Sigma$ converter in CMOS, *IEEE Journal of Solid-State Circuits*, vol. 44, no. 11, pp. 2873–2880, November 2009.

[4] T. Chalvatzis, E. Gagnon, M. Repeta, and S. P. Voinigescu, A low-noise 40-GS/s continuous-time bandpass $\Delta\Sigma$ ADC centered at 2 GHz for direct sampling receivers, *IEEE Journal of Solid-State Circuits*, vol. 42, no. 5, pp. 1065–1075, May 2007.

[5] L. Luh et al., A 4 GHz 4th-order passive LC bandpass delta-sigma modulator with IF at 1.4 GHz, in *Symposium on VLSI Circuits Digest of Technical Papers*, vol. 5, pp. 168–169, February 2006.

[6] J. Harrison et al., An LC bandpass $\Delta\Sigma$ ADC with 70 dB SNDR over 20 MHz bandwidth using CMOS DACs, in *International Solid-State Circuits Conference Digest of Technical Papers*, vol. 5, pp. 146–148, February 2012.

[7] C. Y. Lu et al., A sixth-order 200 MHz IF bandpass sigma-delta modulator with over 68 dB SNDR in 10 MHz bandwidth, *IEEE Journal of Solid-State Circuits*, vol. 45, no. 6, pp. 1122–1136, June 2010.

[8] H. Chae, J. Jeong, G. Manganaro, and M. Flynn, A 12 mW low power continuous-time bandpass $\Delta\Sigma$ modulator with 58 dB SNDR and 24 MHz bandwidth at 200 MHz IF, in *International Solid-State Circuits Conference Digest of Technical Papers*, pp. 148–149, February 2012.

[9] K. Yamamoto, A. C. Carusone, and F. P. Dawson, A delta-sigma modulator with a widely programmable center frequency and 82-dB peak SNDR, *IEEE Journal of Solid-State Circuits*, vol. 43, pp. 1772–1782, July 2008.

[10] T. Shui, R. Schreier, and F. Hudson, Mismatch-shaping for a current-mode multi-bit delta-sigma DAC, *IEEE Journal of Solid-State Circuits*, vol. SC-34, no. 3, pp. 331–338, March 1999.

[11] L. J. Breems, R. Rutten, and G. Wetzker, A cascaded continuous-time $\Sigma\Delta$ modulator with 67-dB dynamic range in 10-MHz bandwidth, *IEEE Journal of Solid-State Circuits*, vol. 39, no. 12, December 2004.

[12] M. Keller et al., On the implicit anti-aliasing feature of continuous-time cascaded sigma–delta modulators, *IEEE Transactions on Circuits and Systems I*, vol. 54, no. 12, December 2007.

[13] R. Bagheri et al., An 800-MHz-6-GHz software-defined wireless receiver in 90-nm CMOS, *IEEE Journal of Solid-State Circuits*, vol. 41, no. 12, pp. 2860–2876, December 2006.

[14] G. Mitteregger et al., A 14b 20 mW 640 MHz CMOS CT $\Delta\Sigma$ ADC with 20 MHz signal bandwidth and 12b ENOB, in *International Solid-State Circuits Conference Digest of Technical Papers*, vol. 1, pp. 62–63, February 2006.

[15] R. Schreier et al., A 375-mW quadrature bandpass $\Delta\Sigma$ ADC with 8.5-MHz BW and 90-dB DR at 44 MHz, *IEEE Journal of Solid-State Circuits*, vol. 41, no. 12, pp. 2632–2640, December 2006.

[16] B. K. Thandri and J. Silva-Martinez, A robust feedforward compensation scheme for multistage operational transconductance amplifiers with no Miller capacitors, *IEEE Journal of Solid-State Circuits*, vol. 38, no. 2, pp. 237–243, February 2003.

[17] A. Thomsen, D. Kasha, and W. Lee, A five stage chopper stabilized instrumentation amplifier, in *Symposium on VLSI Circuits Digest of Technical Papers*, vol. 4, pp. 220–223, 1998.

[18] J. Harrison and N. Weste, A 500 MHz CMOS anti-alias filter using feed-forward op-amps with local common-mode feedback, in *International Solid-State Circuits Conference Digest of Technical Papers*, vol. 4, pp. 132–133, February 2003.

[19] R. Schreier and G. Temes, *Understanding Delta-Sigma Data Converters*. IEEE Press, 2005.

[20] B. Murmann, ADC Performance Survey 1997–2012, <http://www.stanford.edu/~murmann/adcsurvey.html>.

[21] P. Shettigar and S. Pavan, A 15 mW 3.6 GSPS CT-ΔΣ ADC with 36 MHz bandwidth and 83 dB DR in 90 nm CMOS, in *International Solid-State Circuits Conference Digest of Technical Papers*, vol. 4, pp. 156–157, February 2012.

[22] M. Bolatkale, L. J. Breems, R. Rutten, and K. A. A. Makinwa, A 4 GHz continuous-time ΔΣ ADC with 70 dB DR and −74 dBFS THD in 125 MHz BW, *IEEE Journal of Solid-State Circuits*, vol. 46, no. 12, pp. 2857–2866, December 2011.

High-Performance Pipelined ADCs for Wireless Infrastructure Systems

9

Michael Elliott[a] and Boris Murmann[b]

[a]*Analog Devices, Inc., Wilmington, MA 01887, USA*
[b]*Department of Electrical Engineering, Stanford University,*
Stanford, CA 94305, USA

INTRODUCTION

Pipelined A/D converters (ADCs) have been used since the mid 1980s [1–3] to enable wideband digitization at moderate-to-high resolutions of approximately 8–14 effective bits. In the early days of its commercial adoption, the pipelined architecture was primarily used to digitize video signals at approximately 20 MS/s and 10 bits of resolution [4]. At that time, with CMOS feature sizes of several microns, this level of performance was difficult to achieve with competing architectures.

Even though more than two decades have passed, the situation hasn't changed much. Despite the fact that competing architectures (such as successive approximation and oversampling ADCs) have substantially widened their performance space, pipelined converters still enjoy great popularity in high-speed applications. This can be seen from Figure 9.1, which plots experimental ADC data presented at the IEEE International Solid-State Circuits Conference (ISSCC) and the VLSI Circuit Symposium from 1997 until 2012 [5]. For the range of 50–80 dB in signal-to-noise-and-distortion ratio (SNDR), the pipelined architecture dominates the landscape for conversion bandwidths above ∼50 MHz. Within this range, popular applications include wireless LAN [6], Ethernet transceivers [7], medical ultrasound imagers [8], and wireless base transceiver stations (BTS), which are the focus application of this chapter. As we shall discuss in more detail, wireless BTS have in fact become the most significant application that defines and pushes the state-of-the-art in pipelined conversion.

The remainder of this chapter is structured as follows. In Section 9.1, we will investigate the reasons behind the stringent base station performance requirements using a multi-carrier macro BTS as an example. Next, Section 9.2 looks at general issues pertaining to the interface between the ADC and its driving intermediate frequency (IF) section. Section 9.3 describes specific challenges in designs that exclude a dedicated sample-and-hold amplifier (SHA) to save power (and cost). Lastly, in Section 9.4, we summarize recent research that looks into the digital linearization of dynamic nonlinearities at the ADC front-end.

Manganaro: Advances in Analog and RF IC Design. http://dx.doi.org/10.1016/B978-0-12-398326-8.00009-1

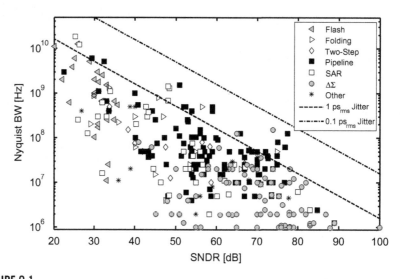

FIGURE 9.1

Published data for ADC Nyquist Bandwidth versus SNDR.

9.1 Receive path specifications for macro BTS

A simplified diagram of a typical macro BTS is shown in Figure 9.2. The receiver section uses a single-stage heterodyne architecture where multiple carriers around the first IF are directly sampled by the ADC. While this approach improves programmability and eliminates I/Q imbalance issues in the second down conversion, it requires the ADC to handle the relatively high IF frequency, which is typically in the range of 100–200 MHz or higher.

Modern wireless standards such as LTE offer high spectral efficiency and relaxed radio performance requirements, but legacy standards such as GSM are still relevant. Today, GSM continues to be supported in most areas due to the historical dominance of the standard. Unfortunately, GSM is also the most challenging standard in terms of radio performance. Thus, to determine the most demanding requirements, it is useful to consider the antenna-referred GSM requirements. For this purpose, Figure 9.3 shows the receiver section with the relevant signal levels at the antenna and ADC input.

The receiver front-end ideally operates at the maximum possible gain to achieve the best sensitivity. The limitation to the gain is the maximum signal level condition. In the revised GSM multi-carrier specification, the maximum signal level at the antenna is – 25 dBm. Unlike the frontend analog components that saturate gracefully with increasing nonlinearity, the ADC presents a hard limit to the maximum signal level because it clips the signal beyond its full-scale range. In a single-carrier GSM BTS clipping can in principle be tolerated, since GSM uses only phase modulation. However, in the multi-carrier case, this is not acceptable since a large blocker that induces clipping would corrupt all of the other channels in the desired band.

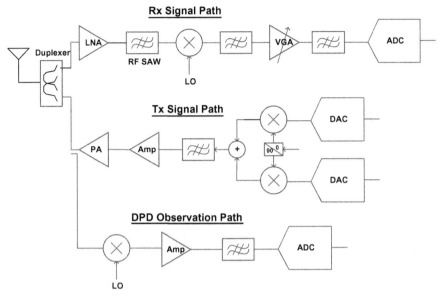

FIGURE 9.2

Multi-carrier IF-sampling BTS system.

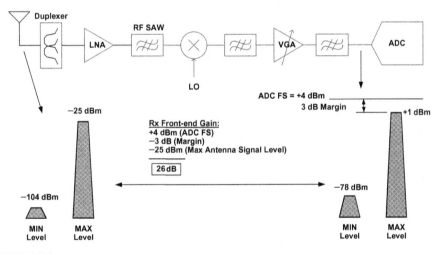

FIGURE 9.3

Receiver architecture and signal levels.

The ADC full-scale range is usually specified in voltage with a typical value of $2\,V_{peak\text{-}peak}$ differential. In the IF section, the differential termination is normally $200\,\Omega$, so the ADC's full-scale corresponds to about $+4\,dBm$. Since clipping is catastrophic, and components have inherent gain variations, some margin is usually built

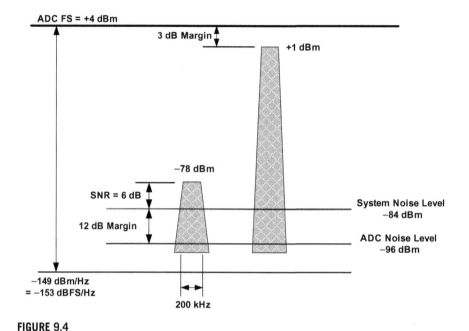

FIGURE 9.4

Signal levels at the ADC input for noise calculation.

into the system. For this example, we assume a 3 dB margin, which results in a maximum gain of 26 dB for the receiver frontend. Note that the radio design can utilize automatic gain control (AGC) to relax the above constraints, but the benefits of that are somewhat limited in a multi-carrier system. We therefore neglect AGC here for simplicity.

Once the signal levels are referred to the input of the ADC, the performance requirements of the ADC can be determined. In Figure 9.4, the minimum signal level is considered to determine the noise specification of the ADC. The signal must maintain a minimum SNR level to allow the carrier to be demodulated, and we assume 6 dB in this example. In a well-designed receiver, the frontend dominates the noise budget. We therefore assume that the ADC noise level is 12 dB below the frontend noise to prevent the added ADC noise from desensitizing the receiver. This leads to a noise power level of -96 dBm, or equivalently a power spectral density of -149 dBm/Hz for the GSM channel bandwidth of 200 kHz. Relative to the ADC's full-scale of $+4$ dBm, the resulting ADC noise specification is -153 dBFS/Hz. In more conventional terms, this translates to an SNR of 72 dB at the standard LTE system sample rate of 245.76 MS/s.

The linearity requirements of the ADC can be computed in a similar manner, by inspecting both the maximum and minimum signal levels, as shown in Figure 9.5. Here, we assume that the margin of the interferer (caused by a strong signal at the maximum level) relative to the minimum signal power is increased to 15 dB.

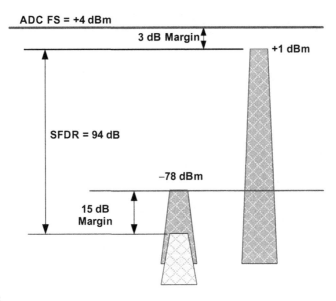

FIGURE 9.5

Signal levels at the ADC input for SFDR calculation.

The total SNDR of the desired weak signal is limited by the integrated power in the channel of both noise and interference. There is some flexibility in this trade-off, but the system considered here is assumed to be limited by noise, rather than distortion. The resulting SFDR requirement of the ADC is 94 dB.

In summary, our calculations lead to receive ADC specifications of SNR $= 72$ dB and SFDR $= 94$ dB at a sample rate of 245.76 MS/s for a typical macro BTS system. These requirements are especially stringent considering that the ADC frontend is sampling IF signals around 150 MHz or higher. That is, the derived SNR and SFDR numbers are harder to achieve at IF frequencies due to clock jitter noise and sampling distortion, both of which increase with input frequency. To get a feel for the required bound on clock jitter, we can invoke the following well-known expression

$$\text{SNR}_{\text{jitter}}[\text{dB}] = -20\log(2\pi f_{\text{in}}\sigma_{\text{jitter}}),$$

which computes the signal-to-jitter-noise for a sinusoid at f_{in}, sampled with a jitter standard deviation of σ_{jitter}. Assuming $\text{SNR}_{\text{jitter}} = 75$ dB to leave some margin, this formula gives $\sigma_{\text{jitter}} = 188$ fs$_{\text{rms}}$, which is achievable only with carefully optimized clock sources and clock distribution networks [9]. This can also be seen from Figure 9.1, which includes $\text{SNR}_{\text{jitter}}$ reference lines for standard deviations of 1 ps and 100 fs. Some of the plotted data points show SNDR performance near the 100 fs $\text{SNR}_{\text{jitter}}$ line. Given that SNDR contains all nonidealities, the clock jitter of these top-performing parts is often below 100 fs$_{\text{rms}}$ (see e.g. [9]).

In addition to the above-discussed receive path, pipelined ADCs also find their application in the digital predistorter (DPD) observation path (see Figure 9.2).

The main challenge here is bandwidth, while the SNR and SFDR requirements are relaxed when compared to the receive path. With a transmit bandwidth of up to 100 MHz in LTE-Advanced applications, the feedback signal bandwidth can be as wide as 500 MHz if the power amplifier distortion has significant fifth-order components. Such systems are currently still under development (see e.g. [10]) and suitable ADC solutions are beginning to appear on the market.

9.2 IF filter and ADC interface

In single-carrier BTS platforms, significant filtering was performed at the IF stage using SAW filters. In multi-carrier systems, the IF filter bandwidth is much wider, which alters the design of the SAW filter. At these wide bandwidths, the SAW filter has a very large insertion loss. This is a significant penalty, so an alternative approach is to utilize LC filters in the IF stage. The LC filters provide some selectivity but the primary purpose is for anti-aliasing prior to sampling in the ADC. Thus, the system's selectivity is almost entirely achieved at the RF stage with cavity and SAW filters, which have low insertion loss at RF frequencies.

The LC filter design begins with the anti-aliasing requirements, as depicted in Figure 9.6. The undesired signals present outside the Nyquist band of interest must be filtered sufficiently to prevent interference as the blockers alias into the signal band. The achievable blocker suppression is controlled by a number of system-level design parameters and conditions such as RF selectivity, ADC sample rate, and adjacent spectral content.

Once the desired filter suppression has been determined, the IF filter transfer function can be selected while accounting for signal band gain variation requirements (filter ripple). When the filter transfer function is set, the only free design variable is the impedance level of the filter. This is a critical design parameter in optimizing the ADC interface, which is shown in more detail in Figure 9.7. A lower impedance level in the LC filter makes it more difficult for VGA to drive its output linearly. On the

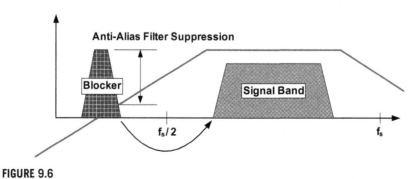

FIGURE 9.6

IF filtering.

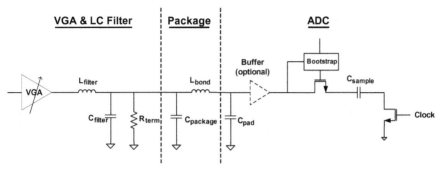

FIGURE 9.7

Interface between the IF section and the ADC.

other hand, choosing higher impedance makes it more difficult for the ADC to sample linearly.

In the simplest case, the ADC's input sampling network consists of a bootstrapped MOS input switch [11], in series with a sampling capacitor and another MOS switch that performs bottom plate sampling [12]. Assuming that the buffer (shown as "optional" in Figure 9.7) is omitted, this switched capacitor network produces significant charge kick-back into its driving network. The corresponding transient glitches interact with the parasitics of the package and the impedance of the LC filter. The final shunt capacitor in the filter absorbs the charge glitch and reduces the resulting voltage glitch. If the nonlinear components of these voltage glitches do not settle prior to the next sampling operation, the linearity of the ADC can be significantly degraded.

There are two basic remedies to this design problem. The first is to integrate a buffer inside the ADC (as shown) to isolate the sampling network from the external environment [9]. This has the advantage that a well-behaved impedance is presented at the ADC input and the charge kick-back is isolated from the circuitry external to the ADC. The disadvantage of this solution is the added power, noise, and cost. The other approach is to try and minimize the charge kick-back and reduce the filter's impedance level, which pushes the problem back to the VGA. This approach yields lower power in the ADC, but these gains may be offset at the system level by the increased power in the driving circuitry. Some of these trade-offs have been analyzed in [13].

The input buffer is typically realized with bipolar transistors, since it proves difficult to achieve the required noise and linearity specifications with MOS transistors. At the same time, the need for higher sample rates in BTS applications favors finer geometry CMOS process technology to efficiently implement the pipeline backend circuitry. As a compromise, many commercial parts use BiCMOS technology, whose CMOS devices unfortunately lag several process generations behind. Ultimately, the process technology selection, frontend architecture, and application space requirements are all intimately connected and a careful cost/benefit analysis is needed to identify the best possible system solution.

9.3 SHA-less front-end design

Once the signal is sampled, the next step is to perform the quantization operation. In a classical pipeline structure, the sampling network discussed in the previous section is used in conjunction with a dedicated sample-and-hold amplifier (SHA) and the quantization is performed by a series of cascaded stages as shown in Figure 9.8. The SHA, however, is a large power and area contributor to the over-all design. By removing the SHA and sampling the input directly with the first pipeline stage, the ADC efficiency is increased significantly [14]. As illustrated in Figure 9.8, for the same noise performance, the power can be reduced by as much as four times.

However, the SHA-less structure is not without disadvantages. The primary issue is that the continuous time input signal now splits into two separate sampling paths: the sub-ADC and the multiplying D/A converter (MDAC) of the first stage. Since the two paths will not perfectly match, one must carefully consider the impact of the error between the two paths. As it turns out, and as explained in [14], the error between the two paths is inconsequential as long as it lies within the overrange volt-age of the stage; in which case the stage's redundancy absorbs the error.

Nonetheless, the error budget is tight when one considers the case of IF sampling with high input frequencies. In this case, sampling errors due to clock skew and/or analog group delay mismatch between the two paths can be significant. As an example, consider the stage realization in Figure 9.9.

Here, the MDAC and sub-ADC signal paths are highlighted to illustrate the critical components that must match for SHA-less operation. In this structure, the switched capacitor network of the comparator implements a replica of the primary sample path in the MDAC. The comparator signal path may be scaled down relative to the

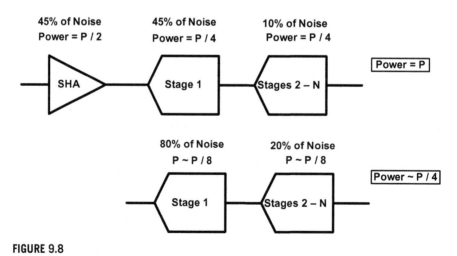

FIGURE 9.8

Pipeline with SHA vs. SHA-less pipeline ADC architecture.

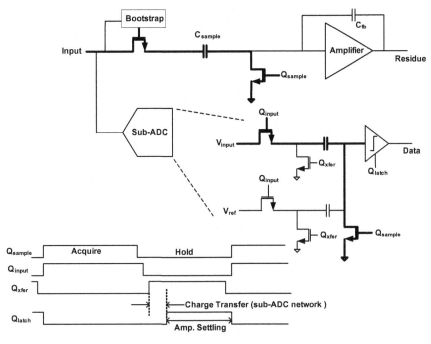

FIGURE 9.9

Sub-ADC in a SHA-less implementation.

MDAC, but the structures are still similar, which improves group delay matching. The penalty with this implementation is that after the sample is taken, the comparator's sampling network must transfer charge prior to latching the comparator. The delay in the comparator latching is restricted by the sampling network bandwidth and the comparator input acquisition bandwidth. The delay in latching the comparator is usually small, but at very high sample rates it results in a significant penalty for the MDAC amplification phase, which now needs higher bandwidth.

The additional delay can be addressed with a structural change to the sampling operation in the sub-ADC. In Figure 9.10, the comparator sampling network was modified to alleviate the propagation delay of the previous structure. In this implementation, the sub-ADC sampling capacitor is pre-charged to the reference voltage during the hold phase (via the clock signal, Q_{ref}). During the acquire phase, the analog input is applied to the capacitor and the comparator tracks the difference of the analog input and reference voltage. The comparator latch core then directly samples the desired signal with the same clock as the MDAC to ensure accurate matching. This structure eliminates the SHA-less structure's charge transfer latency in the comparator decision by removing that operation. The limitation with this approach is the analog bandwidth matching between the comparator latch core and MDAC sample network. The critical factor in this design is the comparator's acquisition bandwidth. If the input bandwidth of the comparator is high, then any mismatch is much less

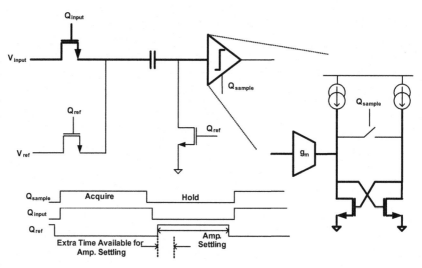

FIGURE 9.10

Modified sub-ADC in a SHA-less implementation.

significant. The MDAC sampling network is operating at very high bandwidths in these applications already to allow for high IF sampling with very flat frequency response, so the comparator is usually the bottleneck.

Consider an example where the frontend must sample at an IF frequency of 500 MHz. The worst case for timing error occurs at the maximum slope of the input signal. For a full-scale sinusoidal input, the maximum slope is

$$\frac{dV}{dt} = 2\pi \cdot f_{in} \cdot V_{fs}.$$

Then, assuming that 5% of the stage's full scale range is available for overranging due to path mismatch errors, the maximum voltage error is

$$dV_{max} = 0.05 \cdot V_{fs}.$$

Combining these expressions, the timing mismatch error budget is

$$dt_{max} = \frac{0.05}{2\pi \cdot f_{in}} = 15.9 \text{ ps}.$$

The skew between the two sampling clock signals is typically smaller than this budget, so the dominant source of error is often the analog group delay mismatch. If the comparator and MDAC both have a single-pole response with a nominal bandwidth of 3 GHz, their group delay is

$$\text{Group Delay} = \tau = 53 \text{ ps}.$$

Thus, a mismatch in the two group delays of about 30% would exceed our total budget of 15.9 ps. Although the circuit structures are not fully matched and their group delay varies with process, temperature, and supply, this level of precision is typically attainable with proper circuit design.

In addition to the above-discussed subtleties in timing, the SHA-less architecture comes with additional problems when applied in a design that also omits the input buffer in Figure 9.7. Specifically, the charge kick-back of certain SHA-less front-ends may contain significant nonlinear components related to the sub-ADC quantization error from the previous sample. The quantization error is a nonlinear function of the input and will therefore cause nonlinear distortion if the input network does not fully settle. To see this, consider the circuit in Figure 9.11, where the sampling capacitors are also used to perform the DAC subtraction.

In this topology, the sampling unit capacitors are charged to the reference voltages during the hold phase. At the transition to the tracking phase, the unit capacitors now hold a charge proportional to the digitized value of the previous sub-ADC sample. This charge is redistributed to the input source impedance when the acquisition phase begins. The input eventually settles to the analog input voltage with sufficient bandwidth in the source impedance and sample network. However, at very high sample rates, this charge does not have sufficient time to settle and the next sample operation acquires a portion of the previous sample's quantization error. The result is a degradation in the ADC's SFDR.

There are several methods to improve this sensitivity to quantization error kick-back. The first is to reset the sample network capacitors briefly prior to the beginning of the acquire phase [15]. This technique reduces the available acquire time and increases the capacitive loading at the input of the ADC due to the additional switches. Another method is to split the sampling network capacitors and the DAC capacitors used for the MDAC operation as shown in Figure 9.12 [14]. This allows the hold phase to perform both charge transfer and the reset operation together. It also disconnects the ADC input from the DAC switches. The trade-off with this structure is degraded noise and increased MDAC amplifier power. The added capacitors

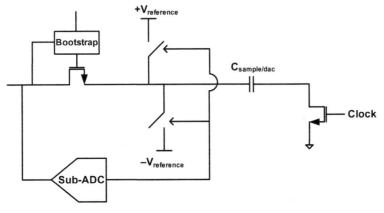

FIGURE 9.11

SHA-less sample network with nonlinear charge kick-back.

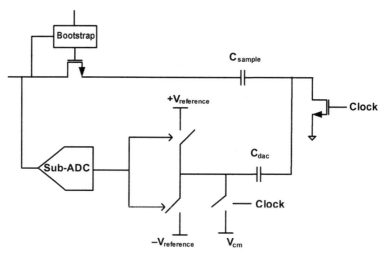

FIGURE 9.12

SHA-less split capacitor implementation to avoid nonlinear charge kick-back.

reduce the amplifier's feedback factor, which degrades the noise performance and adds power to the amplifier to maintain the same bandwidth.

9.4 Switch nonlinearity and digital linearization

As mentioned in previous sections, the requirement of sampling the input signal around a relatively high intermediate frequency results in a number of challenges in the design of the interface circuitry between the system's VGA and the ADC. In this section, we will take a closer look at the distortion due to nonlinearities in the sampling switches. This issue exists in addition to the nonlinear effects related to charge kick-back and is relevant in all of the above-considered circuit topologies (pipeline with SHA, SHA-less, buffered, or unbuffered input).

To investigate, consider the ADC's sampling network during the acquire phase, as shown in Figure 9.13. In this circuit, the switch transistors M1 and M2 are on and the voltage across the sampling capacitor C follows the input signal. At the sampling instant, M2 is turned off first to freeze the total charge at the bottom plate node of the capacitor; this prevents subsequent (signal-dependent) charge injected by M1 from disturbing the acquired sample [12]. When the sampling occurs, the most critical aspect for good high-frequency performance is how linearly the continuous time waveform at V_{in} is mapped into charge at the bottom plate of the capacitor.

There are two main effects that introduce nonlinearities in the sampled charge. The first issue is signal-dependent charge injection and the second problem is tracking nonlinearity. Both of these effects result from input-dependent variations in the on-resistance of M1(R1), which are partly due to backgate effect and finite

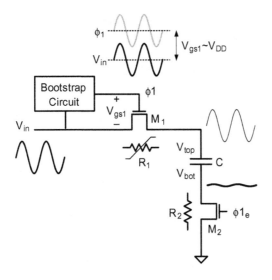

FIGURE 9.13

Sampling circuit in the acquire phase.

bandwidth in the bootstrap circuit. Changes in R1 alter the amount of charge injected by M2 at the sampling instant; this effect typically dominates at low frequencies and is usually not the primary concern in an IF-sampling application. At high frequencies, the main issue is the tracking nonlinearity due to the voltage dependence of R1, which interacts with the sampling capacitor C to introduce frequency-dependent distortion.

The resulting nonlinear relationship with memory between the ADC input and output at the relevant discrete time instances can in principle be expressed as a discrete time Volterra series. Unfortunately, a well-known issue with Volterra models is that even for relatively simple circuits, the corresponding inverse functions (for error correction) can be prohibitively complex. In order to address this issue and to enable efficient digital linearization, the distortion model must be simplified using circuit-specific insight and judicious approximations, as discussed next.

In the circuit's acquire phase, the series combination of R1, R2, and C forms a nonlinear first-order differential equation that links the input voltage (V_{in}) to the capacitor voltage (V_C). Modeling the backgate effect of M1 as the dominant nonlinearity and using first-order device models, we find

$$V_{in} = V_c + (R_1 + R_2)C\frac{dV_c}{dt}$$

and

$$R_1 = R_1(V_{in}) \cong \frac{1}{\mu_n C_{ox}\frac{W}{L}[V_{DD} - V_t(V_{in}, V_{top})]},$$

where we have assumed that M1 is perfectly bootstrapped such that $V_{GS1} = V_{DD}$. In a sampling network with large bandwidth, V_{top} and V_C will track the input closely and differ only by a weakly nonlinear term determined by the respective RC product and the signal derivative. Therefore, the resistance of M1 can be approximated by a nonlinear static function of V_{in}, V_C, and ultimately the digitized output of the pipeline, D_{out}. Taking these considerations into account, the discrete time distortion model simplifies to a product of a memoryless nonlinear function, expressed below as a power series, multiplied into the signal derivative.

$$V_{in}(K) = D_{out}(k) + \left[a_0 + a_1 D_{out}(k) + a_2 D_{out}^2(k) + \cdots \right] \frac{d D_{out}}{dt}(k).$$

This model is considerably less complex than a general Volterra model of comparable order and can be used to compensate for the circuit distortion in the digital domain. Note also that this distortion model is already written in the inverse form, i.e. the ideal input is computed from discrete time samples at the ADC output and no inversion is needed.

One way to extract the derivative of a subsampled IF signal is to interpolate the samples and use a subset of the reconstructed values to approximate the slope. This is schematically illustrated in the schematic of the complete digital post-processor in Figure 9.14. In order to perform the interpolation, the ADC output must be up-sampled and processed by a bandpass filter with a bandwidth of fs/2 centered at the Nyquist zone of the input signal. The bandpass filter is implemented in the digital domain and its required number of taps is chosen based on the accuracy needed in the interpolated samples. In principle, only one reconstructed sample adjacent to the

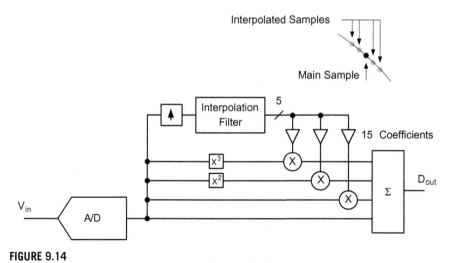

FIGURE 9.14

Digital post-processor for the correction of tracking nonlinearity in the sampling network.

main sample is needed to approximate the derivative. However, due to thermal noise in the ADC and small inaccuracies in the interpolation process, multiple interpolated samples around the sampling point should be used. Furthermore, weighing these interpolated samples independently and multiplying them into the nonlinear power series terms provide additional degrees of freedom for the correction. We found experimentally that this helps absorb second-order effects that are not explicitly contained in the basic distortion model derived above.

As described in more detail in [16–17], we applied the digital processor of Figure 9.14 to a number of commercially available pipelined ADCs targeting BTS applications. In our experiments, a number of single-tone sine waves spanning the desired Nyquist zone were used to determine the optimum coefficients for the post-processor. Figure 9.15 shows SFDR measurements with and without the proposed post-processor in the 5th and 6th Nyquist zones of a 155-MS/s ADC. With digital post-processing, the measured SFDR improves on average by 8–10 dB and remains above 83 dB for inputs up to 470 MHz. The maximum input frequency was limited by our test equipment, and the achieved SFDR was partly impaired by static nonlinearities (INL) in the ADC core.

Whether such digital linearization techniques will find practical adoption in the future will largely depend on the ease and robustness of the coefficient calibration procedure. Since the optimal coefficients are somewhat dependent on temperature [16], they must be measured continuously (or at least frequently) to avoid undesired drift effects. This issue is less significant in test and measurement equipment, where the temperature is controlled within certain limits and a re-calibration can be triggered when large shifts occur. Thus, we are beginning to see dynamic linearization techniques mostly in this application space [18]. For BTS stations, the current trend is still to explore analog circuit improvements, rather than digital correction, to tackle the tracking nonlinearity problem. For instance, the work of [19] explores a dynamically driven deep N-well input sampling switch to mitigate nonlinearities due to the MOS device at the converter input (M1 in Figure 9.13).

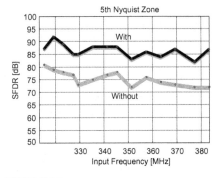 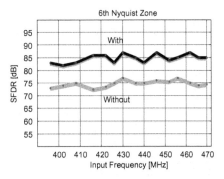

FIGURE 9.15

Measured SFDR improvements with digital tracking nonlinearity compensation.

SUMMARY AND FUTURE OUTLOOK

In this chapter, we have reviewed the performance requirements for pipelined ADCs in the context of their application in wireless BTS systems. Generally speaking, the most challenging issues arise due to the requirement of IF sampling and the associated jitter noise and distortion issues at the ADC input. With proper design, pipelined ADCs remain the only architecture that meets today's receive path requirements with sampling rates around 250 MS/s, SNR > 72 dB, and SFDR > 90 dB for IF inputs.

In the future, the trend of providing increased data rates for mobile applications such as smartphones, tablets, and portable computers will undoubtedly continue. In order to meet these demands, BTS system designs are moving to new communication standards, such as LTE, which are more spectrally efficient. The network design is also evolving into smaller cell sizes to further improve data bandwidth. These changes will in most cases reduce radio performance requirements and impress new power and size constraints. This will enable new technologies to emerge in the radio implementation for BTS such as fully integrated direct conversion and direct RF sampling. In both scenarios, it is expected that high-speed pipelined conversion, as well as most of the design techniques discussed in this chapter, will continue to play an important role.

References

[1] S. Masuda, Y. Itamura, S. Ohya, and M. Kikuchi, A CMOS pipeline algorithmic A/D converter, in *Proceedings of the IEEE Custom Integrated Circuits Conference*, pp. 559–562, 1984.

[2] S. H. Lewis and P. R. Gray, A pipelined 5-Msample/s 9-bit analog-to-digital converter, *IEEE Journal of Solid-State Circuits*, vol. 22, no. 6, pp. 954–961, December 1987.

[3] S. Sutarja and P. R. Gray, A pipelined 13-bit 250-ks/s 5-V analog-to-digital converter, *IEEE Journal of Solid-State Circuits*, vol. 23, no. 6, pp. 1316–1323, 1988.

[4] S. H. Lewis et al., A 10-b 20-Msample/s analog-to-digital converter, *IEEE Journal of Solid-State Circuits*, vol. 27, no. 3, pp. 351–358, March 1992.

[5] B. Murmann, ADC Performance Survey 1997–2012, <http://www.stanford.edu/~murmann/adcsurvey.html>.

[6] S. S. Mehta et al., An 802.11g WLAN SoC, *IEEE Journal of Solid-State Circuits*, vol. 40, no. 12, pp. 2483–2491, December 2005.

[7] C.-C. Hsu et al., An 11b 800 MS/s time-interleaved ADC with digital background calibration, in *2007 IEEE International Solid-State Circuits Conference. Digest of Technical Papers*, pp. 464–465, 2007.

[8] AD9272: Octal LNA/VGA/AAF/ADC and Crosspoint Switch, <http://www.analog.com/en/analog-to-digital-converters/ad-converters/ad9272/products/product.html>.

[9] A. M. A. Ali et al., A 14-bit 125 MS/s IF/RF sampling pipelined ADC with 100 dB SFDR and 50 fs jitter, *IEEE Journal of Solid-State Circuits*, vol. 41, no. 8, pp. 1846–1855, August 2006.

[10] H.-J. Wu et al., A wideband digital pre-distortion platform with 100 MHz instantaneous bandwidth for LTE-advanced applications, in *2012 Workshop on Integrated Nonlinear Microwave and Millimetre-wave Circuits*, pp. 1–3, 2012.

[11] A. M. Abo and P. R. Gray, A 1.5-V, 10-bit, 14.3-MS/s CMOS pipeline analog-to-digital converter, *IEEE Journal of Solid-State Circuits*, vol. 34, no. 5, pp. 599–606, May 1999.

[12] D. G. Haigh and B. Singh, A switching scheme for SC filters which reduces the effect of parasitic capacitances associated with switch control terminals, in *Proceedings of the IEEE International Symposium on Circuits and Systems*, pp. 586–589, 1983.

[13] A. M. A. Ali et al., A 16-bit 250-MS/s IF sampling pipelined ADC with background calibration, *IEEE Journal of Solid-State Circuits*, vol. 45, no. 12, pp. 2602–2612, December 2010.

[14] I. Mehr and L. Singer, A 55-mW, 10-bit, 40-Msample/s Nyquist-rate CMOS ADC, *IEEE Journal of Solid-State Circuits*, vol. 35, no. 3, pp. 318–325, March 2000.

[15] S. Devarajan et al., A 16-bit, 125 MS/s, 385 mW, 78.7 dB SNR CMOS pipeline ADC, *IEEE Journal of Solid-State Circuits*, vol. 44, no. 12, pp. 3305–3313, December 2009.

[16] P. Nikaeen and B. Murmann, Digital correction of dynamic track-and-hold errors providing SFDR > 83 dB up to fin = 470 MHz, in *2008 IEEE Custom Integrated Circuits Conference*, pp. 161–164, 2008.

[17] P. Nikaeen and B. Murmann, Digital compensation of dynamic acquisition errors at the front-end of high-performance A/D converters, *IEEE Journal of Selected Topics in Signal Processing*, vol. 3, no. 3, pp. 499–508, June 2009.

[18] B. Setterberg et al., A 14b 2.5GS/s 8-way-interleaved pipelined ADC with background calibration and digital dynamic linearity correction, in *ISSCC Digital Technical Papers*, 2013.

[19] J. Brunsilius et al., A 16b 80MS/s 100mW 77.6dB SNR CMOS pipeline ADC, in *2011 IEEE International Solid-State Circuits Conference*, pp. 186–188, 2011.

Interleaving of Successive-Approximation Register ADCs in Deep Sub-Micron CMOS Technology

10

Kostas Doris, Erwin Janssen, Yu Lin, Athon Zanikopoulos, and Alessandro Murroni

NXP Semiconductors, High Tech Campus HTC32, 5656AE Eindhoven, The Netherlands

INTRODUCTION

The Successive-Approximation-Register ADC (SAR) architecture receives major attention nowadays because it adapts itself optimally to its deep sub-micron CMOS silicon medium, favoring its simplicity. Its most popular implementation, shown in Figure 10.1, consists of merely a comparator, logic, and a capacitor DAC [1] that approximates serially the input signal. SARs are considered slow converters due to their sequential algorithmic operation, but they cover up this weakness, blending with time-interleaving [2] in ever-increasing numbers, e.g. from 8 units in [3] to 64, 160, and even 320 in [4–6], respectively, and exploiting the speed of advanced CMOS nodes. This combination enables GHz broadband conversion with low to moderate resolutions, good power efficiency, and small area.

The combination of broadband SAR architecture and digital signal processing techniques in deep sub-micron CMOS technologies is today of great interest in cable and satellite receiver applications [7,8], 40 Gb/s and 100 Gb/s optical networks [5,6,9,10], and offers potential for numerous high-data-rate wireless receive applications, e.g. at 60 GHz, UWB, and eBand. With the ongoing scaling of CMOS IC technology, the partitioning over analog and digital is dictated by the optimal implementation of the data converter and its adaptation to the application: circuit functions remain analog only if that is favorable for the overall conversion function from waves to bits. An example can be seen in TV applications, propelled by the increasing demand for higher data throughput over cable networks. Receivers in this domain have evolved dramatically from discrete CAN tuners, to double [11] or single low-IF [12] BiCMOS receivers combined with Surface-Acoustic-Wave (SAW) or Silicon-in-Package (SiP) LC filters, before moving to CMOS zero-IF conversion [13] and finally being realized by CMOS GS/s SARs [7].

Manganaro: Advances in Analog and RF IC Design. http://dx.doi.org/10.1016/B978-0-12-398326-8.00010-8

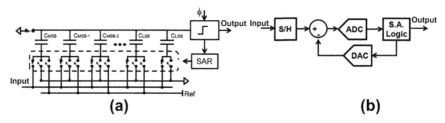

Figure 10.1

(a) Circuit implementation of a charge-redistribution binary SAR ADC [1], (b) generic successive approximation converter.

Broadband sampling with massively time-interleaved GS/s SARs alleviates several issues of conventional receivers related to LO generation and image rejection, oscillator pulling, group delay vs. linearity trade-offs, etc. but introduces issues of its own. In the following sections we review tradeoffs and design degrees of freedom behind these issues and link with representative reception scenarios. In Section 10.1 we briefly review trends in Successive-Approximation (SA)-based conversion for the unit converter in the context of pipelining and interleaving. Next, in Section 10.2 we discuss bandwidth and noise aspects of broadband conversion enabled by many interleaved SARs. Architectural considerations in the signal and clocking paths are discussed in Sections 10.3 and 10.4. Moving on, we focus on the impact of interleaving mismatches highlighting application angles of spectral purity. In the last two sections we present two examples. The first one, in Section 10.6, shows how interleaving many SARs introduces not only issues, but also ways to solve them: in this case redundancy and SAR randomization. The second example, in Section 10.7, demonstrates how broadband sampling introduces new degrees of freedom in spectral sensing for cognitive radio. A summary is given in Section 10.7.

10.1 The amalgam of successive-approximation-based data conversion concepts

Today's SARs are algorithmic melting pots blending together various concepts[1] such as:

- binary [1] or reduced-radix [14] search intervals combined with parallel search [15,16] and pipelining [17,18] with digital [14,17] or analog implementations [4,19–21];
- DAC based on capacitors [1,22,23], resistors [16,17,19], or current [4];
- redundancy in the numbers of ADC units [24,25];

---

[1]The list provided here is not exhaustive. A detailed analysis of the SAR converter goes beyond the purpose of this work.

- dynamic element matching, over-sampling, and $\Sigma\Delta$ modulation [26–28];
- asynchronous logic [22,29,30] that maximizes serial conversion efficiency;
- hardware programming, e.g. of comparator noise during conversion [29,31] or of the conversion resolution [31,32];
- cascaded class-A [4,26] or class-AB amplifications [18,33], charge re-use [29,33], etc.

Time-interleaving and pipelining are two parallel processing techniques possible. Comparisons between interleaved SAR and pipelined ADCs on which architecture is the ultimate one to address high-speed data conversion in modern CMOS processes are popular discussion topics among designers. In this chapter we look at these converters as different implementations of their common successive approximation conversion (SA) algorithm [34,35] because, as can be seen in Figures 10.1 and 10.2, they can always be used orthogonally to each other.

A pipelined converter as shown in Figure 10.2a schedules the operation of N successive approximation hardware units to resolve all bits constituting the digital representation of the sampled input in parallel. Its implementation typically assumes multiple SA units (stages) with additional residue amplification between them. Time-interleaving of generic SA converters is shown in Figure 10.2b. It trades area resources, i.e. number of unit ADCs, with conversion rate [2] and hence makes quantization speed a linear function of the number of units. As a result, it alleviates the need to push amplifiers and comparators to the limits in a given technology node.

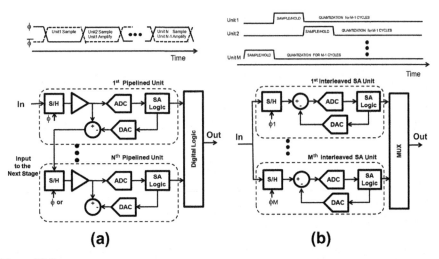

Figure 10.2

Successive approximation conversion with parallel processing: (a) pipeline (bits are processed in parallel), (b) time-interleaved (samples are processed in parallel).

SA converters dominate today's high-speed conversion landscape beyond 100 MS/s at any resolution level.[2] Time-interleaving is more applicable at low to moderate resolutions reaching up to tens of GS/s [6,9]. Deep sub-micron CMOS enables small and high-speed SAR units as well as advanced calibrations that overcome matching, bandwidth, and the jitter limitations stemming from the large SAR array. Traditional pipelined ADCs on the other hand find are widely used at high resolutions, e.g. >12 bit at 100–250 MS/s rates. At these resolutions they are predominantly implemented in BiCMOS processes to make use of bipolar buffers that allow for the handling of large voltage swings and many pF sampling capacitor loads with low noise and high linearity, e.g. [36,37]. Converter evolution taking place in the moderate resolution GS/s performance range (8–12 bit) breaks up the traditional boundaries between these two architectures by combining them. This fusion of concepts offers numerous degrees of freedom for the advance of the conversion function. It can be best understood by the reader by attempting to define a name for an architecture that uses an interleaving hierarchy of 16 two-stage pipelined ADCs, with each stage implemented as an SAR using 2 bit flash ADCs. In the following section we primarily focus at low-to moderate-resolution GS/s interleaved SARs. In this study, pipelining is an additional degree of freedom left to explore alongside many others.

10.2 Bandwidth and noise considerations

Interleaved SARs are associated with the potential to achieve broadband sampling with low noise spectral densities (NSD) and good power efficiency. The ability to do so is determined by bandwidth, thermal, quantization, and jitter noise performance. The relative importance between these varies with the signal properties defined by the application, and by the implementation that determines how these map to circuit mechanisms.

Insufficient bandwidth translates to SNR limitations at high frequencies. For the broadband signal shown in Figure 10.3a signal losses translate to a smooth SNR reduction, with the need for bandwidth increasing when second Nyquist zone sampling (sub-sampling) is considered, e.g. Figure 10.3b. Limited bandwidth in composite signals such as the ones depicted in Figure 10.3c and d is more harmful. For example, sampling a multi-carrier TV signal as in Figure 10.3c with 3 dB bandwidth at the edge of the 1 GHz TV band translates to 3 dB SNR loss for the channel at that location. Similarly, the 5 MHz channels of a sub-sampled multi-carrier WCDMA at near 2 GHz, e.g. Figure 10.3c, will suffer from bandwidth limitations. Consequently, bandwidth requirements for these signals expressed as a ratio over their carrier frequencies are very high. The requirements become even higher when bandwidth matching is taken into account [38].

---

[2]This is a rather arbitrary choice reflecting the interpretation of high speed nowadays, and certainly evolves with time.

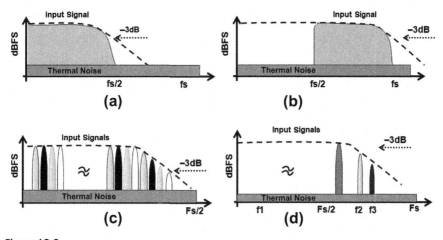

Figure 10.3

Bandwidth impact for various signal types: (a) broadband signal, (b) broadband bandpass signal at second Nyquist band, (c) multi-carrier composite signal loading the Nyquist band, (d) multi-carrier composite signal at the second Nyquist band.

In multi-carrier applications, the need for bandwidth is accompanied by a similar need for low NSD to enable sufficient receiver sensitivity, e.g. below $-150\,\mathrm{dBFS}/$ Hz [39]. The traditional way to achieve high SNR using large signal swings and low intrinsic thermal noise by use of large capacitors and sufficient power is not a straightforward way for the SAR in new generations of CMOS processes due to the reduced supply levels and speed limitations stemming from large capacitors. Interleaved SARs offer another opportunity to improve SNR leveraging their simple architectures with few noise sources and exploiting CMOS transistor speeds to achieve large over-sampling gain. Fundamentally, the tradeoff between thermal noise limited SNR and sampling rate follows a 10 dB per decade relationship [40]. Low NSD at high frequencies can thus be achieved with very high sampling rate, reflecting a tradeoff between achieving low instrinsic thermal noise in each sample, or by sample-to-sample averaging. This is exploited in the A/D converter itself, e.g. in high-resolution SARs [26–28] and in the receiver functionality, e.g. the gain for a 6 MHz TV channel at 2.6 GS/s is more than 23 dB [7].

The real noise limitation, however, at high frequencies is timing jitter. In contrast to thermal noise, artifacts induced by jitter and other timing errors are highly correlated to the input signal. Sampling errors are generally described as a multiplication of the *signal slope* with the *timing error*. The slope of the signal encapsulates application properties, since it relates to the signal statistics, to its narrowband or broadband power spectral density and crest-factor, to its multi- or single-carrier composition, etc. The timing error encapsulates circuit properties, e.g. thermal noise in the clock distribution and samplings circuits translated into white Gaussian timing jitter, phase noise in the PLL, mismatch-based timing errors between interleaved

SARs. Analysis and literature on the impact of jitter in A/D and D/A conversion and on sampling jitter in broader signal processing contexts can be found, for example, in [41–49], and their references.

White Gaussian timing jitter generates broadband noise irrespective of the input signal. The correlation with it is reflected in the dependency of the NSD with the signal power and frequency allocation relative to the sampling rate. The noise level increases as the signal frequencies get higher, making it dominant at high frequencies. For a single sinusoid, a ten-fold increase in frequency translates to a 20 dB increase in NSD. Over-sampling also offers noise spreading benefits reducing the NSD penalty to 10 dB/dec when both the signal and the sampling rate are increased and the actual timing error stays constant. This suggests again that low intrinsic noise can be traded with a high sampling rate (averaging).

The impact of timing jitter and phase noise is conceptually shown in Figure 10.4 for four representative cases. Bandlimited random signals occupying large portions of the Nyquist band such as the ones shown in Figure 10.4a and b are well known to suffer less from white Gaussian jitter compared to sinusoids at maximum frequency due to their smaller derivatives. Moreover, they are less sensitive to low-pass power losses and benefit the least from the small available over-sampling. In the composite multi-carrier signal in Figure 10.4c each individual carrier signal is affected by the noise generated from all the rest, and by its adjacent neighbor carriers in the case of close-in phase noise. This is counterbalanced by the smaller derivative of the composite broadband signal and the large over-sampling gain per carrier.

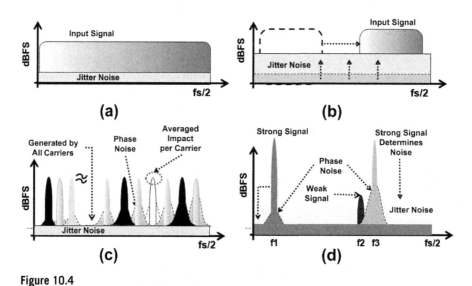

Figure 10.4

Conceptual impact of jitter and phase noise in a (a) broadband signal, (b) broadband bandpass signal, (c) multi-carrier signal with fully occupied Nyquist band, (d) multi-carrier signal with partially occupied Nyquist band.

Power variations eventually lead to tighter jitter specifications since specifications are always set by the weakest signals. Multi-carrier signals as in Figure 10.4d are affected significantly by the large power variations between their sub-carriers. Signal concentration at specific frequencies takes away the advantage of small derivatives whereas large signals at high frequencies determine here the jitter noise level. In the example shown in Figure 10.4d with 100 fs white uncorrelated jitter, a −10 dBFS signal located at 2 GHz generates a NSD of around −160 dBFS/Hz.

How easily intrinsic amplitude and timing accuracy (large signals and large capacitors, high resolution, low timing jitter) can be traded with over-sampling depends on the power efficiency of the converter at high speeds, which is a function of CMOS evolution and circuit innovations. Speed improvements in $\Sigma\Delta$ modulation, e.g. [50], indicate the break-even point is still below 100 MHz Nyquist bandwidth. For conversion beyond 100 MHz, the main enabler of good SNR remains low intrinsic thermal noise in both amplitude and time domains although speed and power efficiency improvements in interleaved SARs keep pushing this boundary. Comparing the NSD achieved in [36,25,9] at 100 MS/s, 1 GS/s, and 10 GS/s, respectively, we observe −156 dBFS/Hz at 125 MHz [36], −144 dBFS/Hz at 1 GHz [25], and −132 dBFS/Hz at 10 GHz [9], suggesting a drop greater than 10 dB/decade with signal frequency and sampling rate for comparable power consumption.

Finally, the choice of receiver architecture can also influence the actual needs for NSD. Broadband sampling can replace zero-IF down-conversion stages with direct- or sub-sampling and permits the GS/s SAR to claim more noise budget via the removal of mixers, filters, image rejection effects, frequency synthesis spurs, etc. For example, the zero-IF multi-channel receiver in [13] employed additional multiple 11 bit ADCs at 175 MS/s with 200 mW compared to the use of a single 500 mW 10 bit 2.6 GS/s ADC with the direct sampling architecture in [7].

10.3 Interleaved SAR architecture overview and tradeoffs

Interleaved SAR architectures reflect the design issues stemming from the large number of converters that share the same input, timing, and amplitude reference signals. The key considerations involved are graphically depicted in Figure 10.5 and can be broadly defined as *capacitance* and *matching* management issues. Managing the signal and clock distribution capacitance with appropriate buffering over-shadows the bandwidth, noise, and linearity potential of the unit SAR at high frequencies; timing, bandwidth, gain, and offset matching limitations between the units define spectral purity. The SAR unit dimensions and speed, the CMOS technology generation, and various design degrees of freedom form the scaling factor of the aforementioned issues. Signal and clocking architectural considerations will be reviewed in Sections 10.2 and 10.4, respectively. The impact of interleaving mismatches on representative cases of signals is discussed in Section 10.5.

CMOS technology scaling reduces the unit SAR area, thanks to finer geometries and calibrations that remove matching area considerations, and it increases the speed

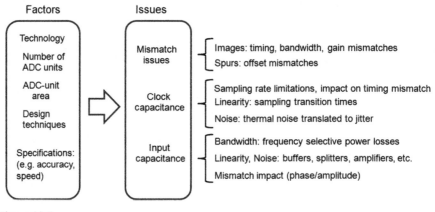

Figure 10.5

Capacitance and matching considerations in time-interleaved converters.

of the unit through raw speed of devices. These help reduce the interconnect capacitance from the input, enabling the architecture shown in Figure 10.6a. Here, the input signal from the source is distributed directly to the bank of capacitors of the charge redistribution ADC unit [1]. Noise stems from one sampling and one quantization operation, rendering this approach very efficient when the ADC meets thermal noise requirements only. Linearity is limited only by the switched capacitor sampler, the static characteristic of the DAC, and the reference buffer. The ADCs in [3,51] are representative examples of this architecture. The ADC in [51] interleaved 24 SAR ADCs reaching 2.8 GS/s with thermal noise levels close to 50 dB below full scale with low power consumption and good linearity. Jitter and bandwidth limitations close to and beyond 1 GHz suggested that small unit size alone is not sufficient for high SNR and linearity at high frequencies.

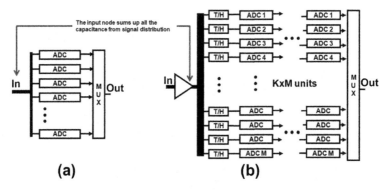

Figure 10.6

Interleaved architectures: (a) small array without buffering, (b) large array with re-sampling, pipelining, and input buffer.

The key challenge with many units and large signal and clock distribution area is hiding the impact of interconnect without noise and linearity penalties. The straightforward approach is to use an input buffer and pipelining [52], shown conceptually in Figure 10.6b. In its time, this 0.18 μm CMOS ADC pushed interleaving to 20 GS/s at 8 bit, combining 80 current mode pipelined ADCs at 250 MS/s each, and dissipating 10 W in total. The SiGe buffer dissipated 1 W. The 6.6 GHz 3 dB input bandwidth—below the Nyquist rate—and 29.5 dB SNR at 6 GHz limited by jitter highlighted the relevance of signal and clock distribution over large areas.

Another way to handle the input interconnect capacitance is to absorb part of it into the ADC using pipelining and hierarchy to reduce the dimensions and numbers of T/Hs. Recent examples in the literature combine pipelining [17], with sampling hierarchy [5,9], and with sampling hierarchy and multiplexing [4]. Figure 10.7a shows the first approach, wherein an additional T/H drives each ADC with an open-loop interfacing buffer. The physical separation between T/Hs and ADCs shrinks the input signal and clock distribution network, part of which is now absorbed by the driving capability of the buffer. The larger conversion rate of the pipelined SAR compared to a single SAR reduces the number of T/Hs and SARs, at the cost of need for linear interstage amplification. For the 0.13 μm CMOS ADC in [17] the ratio between T/H and ADC arrays with 16 units each was around 11, whereas pipelining further enabled reaching 1.8 GS/s.

Sampling hierarchy extends pipelining of the sampling operation further. The basic idea is to bring down the sampling operation from a few T/Hs to multiple ADCs hierarchically. In Figure 10.7b this is depicted for the general case of K sampling stages. For two stages, combining M T/Hs, each driving L unit ADCs with an interfacing buffer, the aggregate rate of the converter becomes $M \cdot L$, whereas clocking and input distribution complexity reduces to that of M units. Hierarchy in the signal

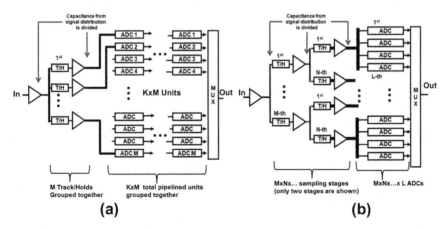

Figure 10.7

Interleaved architectures: (a) with re-sampling and pipelining, (b) with sampling hierarchy and input buffer.

distribution can also be further combined with power splitters [5,9] and binary buffer trees after the T/H [53].

Sampling with or without hierarchy as in Figure 10.7a and b trades less interconnect loading at the input source with more sampling capacitance and more noise: for the same SNR specification, compared to the approach in Figure 10.6a both the input source and the interfacing buffer are loaded with a sampling capacitor that is four times larger due to the impact of re-sampling noise. For thermal noise limited ADCs at low resolutions the sampling capacitor is significantly less than the input interconnect load and the benefits are significant. At high resolutions the situation inverses. The sampling capacitor is dominant and re-sampling has a large impact on performance, therefore this approach is not attractive. In the range between 8 and 12 bits, interconnect and sampling capacitors can be comparable to each other, suggesting careful consideration of all options.

Two-stage sampling hierarchy for 6 bit charge redistribution SARs was demonstrated in the 65 nm and 45 nm ADCs in [5,9], respectively. Aiming at optical networking applications with broadband signals, these ADCs combined a total of 160 unit converters in 16 groups of 10 units each reaching 40 GS/s rate. Power splitting at the input helped to enhance bandwidth, but once more, its use is prohibitive at high or even moderate resolutions. This ADC achieved 25.2 dB SNDR at 18 GHz dominated by 0.25 ps jitter, timing skew, and high-frequency nonlinearity.

Broadband T/Hs realized with the combination of a (bootstrapped) sampling switch and capacitor, and a source follower or unity gain amplifier [54–58,17,59,60,5,9] are keys to good SNR performance at GHz frequencies. However, this combination prohibits achieving simultaneously high linearity and SNR above 50 dB levels near or beyond 1 GHz frequencies. The key considerations are depicted in Figure 10.8:

- Sampling at GHz frequencies with high linearity requires considerations traditionally placed at high-resolution converters [36,61] and constrained by GS/s rates. These are related to switch nonlinearity, impedance matching, supply, and common mode disturbances, etc.

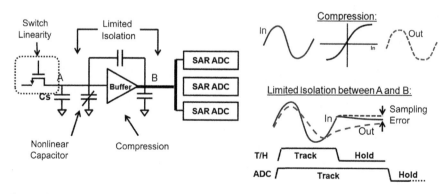

Figure 10.8

Nonlinearity mechanisms in hierarchical sampling.

- The compressive transfer function between input and output nodes A and B and the output resistance modulation of the buffer introduces a noise–linearity–speed tradeoff partially addressed by replica well-bias [58,59], active bootstrapping [17,62], cascoding [56], use of higher supplies [9,17], and digital corrections [63].
- A large buffer enables reduction of capacitance at input and clock nodes but requires large sampling capacitors beyond thermal noise needs to linearize its nonlinear gate load. Thus, less interconnect at the input at the cost of more sampling capacitance.
- Finite isolation between input and output of the interfacing buffer removes charge from the floating sampling node during re-sampling [17,25], bringing tradeoffs between the main sampling capacitor, the buffer's load, its slew rate and bandwidth, the signal amplitude, and the available tracking time.

Hierarchical sampling and higher resolution make these tradeoffs more difficult to address due to the increased interconnect and sampling capacitors the buffer needs to handle, resulting in stringent slew rate and bandwidth requirements. These necessitate more power, higher supplies, and simultaneous tracking of the input signal with many T/Hs [64]. The latter trades relaxed slew rate with additional sampling capacitor load at the input. Once more, at low resolutions the full advantages can be exploited but as the resolution increases this freedom is restricted.

The combination of hierarchical sampling with multiplexing in the 65 nm ADC [4] introduced two additional degrees of freedom. As shown in Figure 10.9a a two-stage sampling hierarchy combining four T/Hs driving 16 current mode SAR ADCs each scaled down the input interconnect dimensions. This was followed by a second

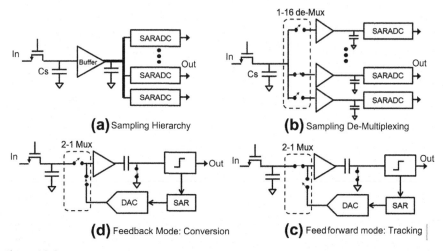

Figure 10.9

Hierarchical sampling in [4] (a) hierarchy, (b) de-multiplexing, (c) feed-forward mode, (d) conversion feedback mode.

step partitioning the interface buffer in to 16 smaller ones, one for each SAR as in Figure 10.9b. A de-multiplexer placed between the front-end T/H and the buffer array decimated the signal distribution capacitance of the 16 ADCs that would otherwise load the main sampling node, and enabled small buffers driving 10 bit thermal noise load. In the last step shown in Figure 10.9c and d the buffer was introduced in the SAR loop to avoid compression limitations but still benefit from its speed. These techniques helped reaching an SNR of 54 dB at lower frequencies (thermal noise limited), and 50 dB up to 4 GHz (jitter noise limited), with total harmonic distortion levels lower than −55 dB up to 2 GHz. Furthermore, it allowed using redundant SAR units as it will be shown in Section 10.6.

10.4 Clocking considerations

So far, we have covered aspects related to the signal path of the converter determining thermal noise and linearity performance. The clocking path defines jitter noise and timing error mismatches and has a major impact on the power consumption of the converter. The role of the sampling load is now represented by the inherent gate load of the sampling switch that is driven by the clock source, while the input distribution interconnect is replaced by the clock, which is influenced by the number and dimensions of the T/H and SAR arrays.

The hierarchical sampling and multiplexing concepts presented in Section 10.2 reduce the dimensions and the number of T/Hs that require precise clocking, simplifying clocking as a whole: less clock cascading (thus, less jitter accumulation), steeper edges due to less clock distribution capacitance, less power consumption, etc. Similar concepts are also applied at the clock path to manage sampling rate, noise, and matching tradeoffs. When the T/H array is small a few hundreds of femtoseconds of timing accuracy can be reached with intrinsic design techniques. Sampling switch sharing [65,66] uses a multiplexer in series with a global sampling switch and the sampling capacitor, and thus suffers from bandwidth mismatches. Clock de-multiplexing or clock gating, e.g. [4,17,51,67], makes one edge of the main clock available to a single clock de-mux unit in the bootstrap circuit, which then determines the timing accuracy. Since large sampling switches driven with fast sampling transients have a chain effect at the clock loading, especially with many T/Hs, this approach also helps to reduce gate loading. Susceptibility to process, temperature, and voltage variability are its main drawbacks.

Many interleaved T/Hs require hierarchical clock generation and calibrations based on multi-phase clock generators and high-frequency clock dividers, clock interpolators, and a choice between CMOS or Current Mode Logic dependent on power consumption and interference susceptibility considerations. Transmission line effects, electromagnetic coupling, current return paths, etc. are essential elements to analyze for a properly designed clock distribution network.

Data collected in surveys [68] indicate a slow but constant improvement of intrinsic jitter performance in recent years, with today's high-speed interleaved converters reaching close to $100\,fs$ levels and not far away from non interleaved state-of-the-art ADCs that are less restricted for low timing jitter. Jitter improvements for inter-leaved SARs build on the tradeoffs mentioned so far. On the one hand, clock signal slopes increase with (a) better matching and higher intrinsic speed of devices in deep sub-micron CMOS, (b) the capacitive load reduction stemming from finer geom-etries and less local interconnect between clock units, and (c) faster and smaller SAR units, thanks to calibrations. This reduces the translation of thermal noise to jitter and mismatch to timing skew. On the other hand, the reduced supply levels, resulting from CMOS evolution and the tendency to interleave more, reduce slopes, and hence increase timing errors. The net result is a slowly improving intrinsic jitter level, how-ever, it is accompanied by a constant improvement of conversion rates that delivers improved NSD due to noise spreading.

10.5 Interleave mismatches

Timing, bandwidth, gain, and offset mismatches between the unit SARs form the main obstacle for the widespread acceptance of interleaved converters in applications where spectral purity is of primary importance. Their impact has been extensively analyzed in the literature, e.g. [38,69,70] for single sinusoids. In this section we look at their relative impact on signal scenarios requiring high spectral purity and we give an overview of existing calibration practices.

For a time-interleaved ADC with M units, gain, timing, and bandwidth mis-matches create *images* of the input signal at the ADC output at frequencies deter-mined by [38]:

$$f_k = f_{in} + \frac{k}{M} \cdot f_s, \quad k = 1, 2, \ldots, M - 1. \tag{10.1}$$

Moreover, it is useful to remember that:

- Gain errors are amplitude modulation errors. The resulting image power depends on the magnitude of the gain error on the input signal, but not on its frequency.
- Timing errors translate to signal errors via the derivative of the input signal.
- Bandwidth mismatch has two parts: gain and phase, with the latter being usually dominant below the $3\,dB$ bandwidth. The gain mismatch in this case is frequency dependent.
- The image amplitudes depend strongly on the distribution of the timing, gain, and bandwidth error among the units.
- The number of images and error power distribution among them is proportional to M.

Offset mismatches cause fixed spurs independent of the input signal. The spurs are located at frequencies determined by the sampling rate and the number of interleaved units:

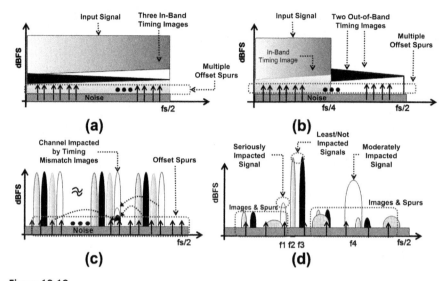

Figure 10.10

Interleave mismatch impact in: (a) broadband signals, (b) influence of over-sampling, (c) multi-carrier signal with fully occupied Nyquist band, (d) multi-carrier signal with partially occupied Nyquist band.

$$f_k = \frac{k}{M} \cdot f_s, \quad k = 1, 2, \ldots, \lfloor M/2 \rfloor. \tag{10.2}$$

Let us first consider signals with power evenly distributed over the Nyquist band as in Figure 10.10a and b, representing sampling scenarios for short-range communication receivers, e.g. 60 GHz, optical communications [6,10], etc. Figure 10.10a shows three timing[3] images generated when sampling such a signal with four interleaved T/Hs. The three images indicated in the figure have frequency shaping following their dependency with the input signal and they are assumed to have different power. Their error power adds together (not shown in the figure). Offset spurs in this example are generated by a large number of unit ADCs.

For this signal, the signal to total error power ratio stemming from interleave errors within the Nyquist band matters, i.e. the impact is averaged over the Nyquist band. For timing and bandwidth mismatches the ratio of the signal power to the image error power is significantly less than that of a sinusoid at Nyquist frequency thanks to its smaller derivative (e.g. up to three times smaller white uncorrelated bandlimited signals [71]). Over-sampling spreads interleaving error power away

---

[3]Bandwidth mismatches result in similar spectra.

from the band of interest, hence improving the total signal to error power ratio. As is shown in Figure 10.10b, for a four times interleaved T/H with $f_s > 4 \cdot BW$, two out of three images of the input signal are out of band, translating to a factor 3 improvement.

A composite signal such as the one shown in Figure 10.10c consists of many narrow-band components. Each channel bandwidth can be substantially smaller than the ADC's sampling rate, but the composite signal can be comparable to it. The high number of modulated carriers, e.g. up to 150 channels with 6 MHz each in TV applications, generates significant interferences due to interleave mismatches. The important parameter here is the signal to total error power ratio within the weakest narrowband channel. The main considerations are:

- Timing, gain, and bandwidth images are evenly averaged per channel, similarly to the broadband signal. Only those aggressor channels that relate via Eq. (10.1) with a victim channel superimpose image power to it, weighted by the aggressor's frequency (for timing/bandwidth).
- Concentrated offset spur power has a much larger impact at the narrowband signal. The SFDR of the spur and not the total average spur power is important.
- The more the interleaved units, the smaller the concentration of offset power and hence the impact to a narrowband channel.
- The larger the signal differences between channels, the larger the impact.
- Over-sampling spreads mismatch power, however, specifications are still set by those few channels that are still maximally affected.

The peak to average power ratio (PAPR) of the signal increases the relative impact of offset mismatch compared to timing, gain, or bandwidth mismatch proportionally to the back-off factor applied. This suggests an impact of up to 12 dB for signals with Gaussian amplitude distributions. In addition, the power spread among interleaving spurs or images follows the actual physical mismatch origin, e.g. threshold voltage spread in clock buffers and switches, or in comparators and amplifiers. This can easily lead to 4–5 dB variations in spurious or image power though the average power stays constant. On top of that, the specification for each interleave error type needs to be carefully balanced together with other error sources for the weakest channel.

The discussion so far suggests that direct translation of specifications based on single sinusoids at maximum frequency may lead to over-specification for timing and bandwidth mismatch, and to under-specification for offset mismatch. For the cable TV example considered here, typical requirements for timing skew are <500 fs rms whereas for offset spurs it should be below −80 dBFS. These values highlight the challenge of designing on-chip calibrations that guarantee these values under all process, supply, and temperature corners with a large enough yield. An example of the impact of all these effects with measurements can be seen in Figure. 10.11. This figure is based on real measurement data using the receiver in [7] in a complete cable TV application set-up. The input signal consists of 150

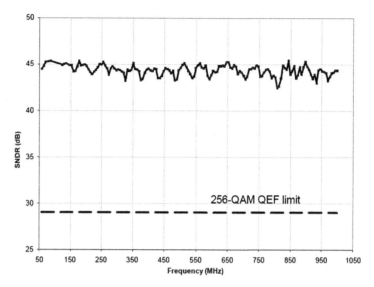

Figure 10.11

Signal to noise and distortion ratio of one QAM channel in a fully loaded composite multi-carrier cable TV signal.

sinusoidal carriers and one QAM modulated carrier. The SNDR of the latter is measured over frequency and contains the combined impact of all ADC's imperfections. The SNDR per channel is dominated by thermal noise with a clear impact from offset spurs that introduce near 2 dB reduction, in accordance with our earlier discussion.

The impact of images and spurs in a generic wireless reception scenario, e.g. broadcast TV or cellular, is conceptually shown in Figure 10.10d. The key considerations are:

- Images from timing, bandwidth, and gain errors are concentrated in the power spectrum in accordance to their correlation with the input signal, consisting of multiple narrowband components sparsely occupying the GHz spectrum.
- The impact depends on the interplay between carrier frequencies, bandwidth and power of each band/channel, the sampling rate, and the number of units.
- Large power differences between signals may lead to very challenging specifications, especially at high frequencies due to timing/bandwidth mismatch dependency with frequency.
- The impact of offset is similar to that of the composite multi-carrier signal in Figure 10.10c, though they can be easier avoided by proper choice of sampling rate and number of units.

Typical wireless sensitivity scenarios assume ADCs with noise floors below −150 dBFS/Hz, which translate to an SNR of near 60 dB for a full-scale sinusoid sampled at 3.2 GS/s. For an assumed 5 MHz wide channel this results in a total thermal noise level of approximately −85 dBFS. Therefore any offset spur that co-resides with this input signal must be well below this value to avoid impacting the sensitivity of the receiver. Taking into account implementation margins the SFDR specification can easily reach values of −90 dBFS.

Sub-sampling wireless signals with broadband GS/s converters is an example that demonstrates how challenging timing error mismatch can be. Suppose that a channel with −75 dBFS is sampled together with an aggressor signal at −10 dBFS at 10 MHz frequency offset from each other within the WCDMA Band-I at 1950 MHz using second Nyquist zone sub-sampling. With a Gaussian timing error spread of 0.1 ps rms and four T/Hs generating three tones, the spurious power per tone can easily reach −60 dB below the aggressor and thus fully block that channel. Proper choice of (a large enough) sampling rate offers here the possibility to selectively avoid interfering images and remove them easily with embedded digital filtering operations.

From the discussion so far, it should not be surprising to the reader how relevant mismatch power is, especially for multi-carrier applications at GHz frequencies. As a result, power efficient mismatch calibrations is one of the main enablers of GS/s SAR conversion. Today's ADCs use background and foreground calibrations with digital and analog detections and corrections (actuation), on-chip digital implementations of algorithms, on-chip signal generators, memories, etc. It is not uncommon for the calibration engine to occupy nearly as much or even more area than the SAR array itself [9,25].

Offset errors are usually assumed to be the easiest to solve because they can be detected by fairly easy averaging operations. This is true when the only objective is to have low average power. However, we have seen that offset spur requirements can be extremely low even for 10 bit converters. At this level, 1/f noise contributions of amplifiers, comparators and their offset actuators, quality and stability of the on-chip reference signal, amplitude and stability of the input signal (when it is used as reference), averaging time, and the large number of interleaved units introduce substantial calibration complexity. Next to calibrations, traditional offset cancelation techniques offer additional means to deal with such offset levels [72], but they are not the only ones at the designer's disposal, as shown in the next section.

Timing error detection methods make use of off-chip or on-chip FFTs [5,52], zero-crossing detection [73,74], cross-correlations [63], etc. The ADC in [9] included an on-chip high-frequency sinusoidal signal generator to be used as reference for timing skew measurements, whereas the ADC in [63] employed on-chip DAC with random signal generation. Timing error correction is done in both analog and digital domains based on various implementations of adjustable delays, e.g. [9,52,63,74,75], or digital equalizations [76]. The impact of bandwidth mismatch is often reduced by designing for a large tracking bandwidth. However, for large sampling capacitors at high resolutions this approach is very challenging to realize. The ADC in [75] used supply regulation to tune the bandwidth of the sampling switch. Independent of the method, there is hardly any work today reporting timing accuracy below approximately 100 fs. This limitation is primarily partially attributed to the difficulty distinguishing accurately the interaction between gain, timing, and bandwidth errors.

10.6 **SAR randomization techniques**

At first sight it might seem that the spurious SFDR, and thus the SNR in a channel of a multi-carrier signal, can only be improved by reducing the total error power by means of calibrations. Interleaved SARs, with the help of their great number of units, offer means to influence the concentration of interleave error power:

- use of a large number of SAR units to spread gain/offset error power;
- separation of interleave errors of the T/H and SAR units thanks to sampling hierarchy;
- using redundancy in the number of T/H or SAR units and randomizing their operation sequence to spread error power over the complete frequency range.

In a traditional time-interleaved converter the SAR units are operated in a fixed (sequential) order, which results in a repeating pattern of time-interleaving errors and the associated fixed frequency tones. By integrating more units than required to achieve the sampling rate, it becomes possible to dynamically change the order in which they are operated, hence breaking the periodicity of the error pattern. If enough redundant units are available to randomize the operation order, the error tones will be transformed in wide "bumps" of noise. This operation only distributes the error power more evenly over the frequencies causing an SNR improvement for the channels that before randomization were polluted with an offset tone, and it will slightly reduce the SNR for the other channels, since these now also see part of the offset error power. Since the performance of the complete receiver is determined by the performance of the worst channel, this technique can typically improve the overall performance by several dBs. Obviously, the penalty for this improvement is an increase in silicon area to accommodate redundancy.

In the literature several publications can be found that describe simulation results of basic randomization [77–81], but only a few report measurements [82,83]. The reported measurements show that the addition of one redundant ADC unit can have a significant effect on the performance of the converter. In the remainder of this section measurement data obtained on the basis of [25] will be detailed.

The time-interleaved converter of [25] consists of four groups of 16 SAR units. Since only 12 units are required per group to achieve the sampling rate, there are 4 redundant ones that can be used for randomization. The effect of enabling the randomization process can be seen by comparing Figures 10.12a and b, which show the measured output spectra of a single quarter producing 650 MS/s, without and with randomization, respectively, for a 111 MHz full-scale input. In both cases 16 ADCs are used for interleaving. In Figure 10.12a we notice a multitude of spurs, resulting

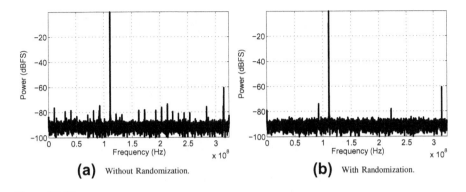

(a) Without Randomization.　　　　　**(b)** With Randomization.

Figure 10.12

Spectral plot with randomization (a) disabled, and (b) enabled using 16 SAR units.

from gain and offset mismatches between the 16 units translated to noise in Figure 10.12b. Because the input signal is at full scale, in both cases the SFDR is limited by the harmonic distortion to 60.5 dB. Randomization has no impact on the SNDR, which measures 53.7 dB in both cases. If we exclude the harmonics, the SFDR is 73.6 dB without randomization and 84.5 dB with randomization, i.e. an improvement of almost 11 dB. From this example it is clear that randomization is an effective method for turning discrete interleaving spurs into wideband noise, thereby improving the conversion quality of time-interleaved converters, especially when used to receive a multitude of narrow-band carriers.

10.7 **Spectral sensing with broadband sampling**

This section presents an example of spectral sensing enabled by integrated broadband interleaved SARs. The successful operation of a cognitive radio (CR) strongly depends on its capability to do spectral sensing. The process should not only be very fast, but it should also be very sensitive. Fast sensing is required in order to locate an unused frequency band in fractions of a second, and the high sensitivity is required to detect weak signals with an SNR as low as −20 dB in order to avoid interference [84].

Although the CR that can work over several GHz of bandwidth is still far from reality, the IEEE 802.22 Wireless Regional Area Network (WRAN) standard is based on CR technology in the television broadcast band [85]. In this standard the wireless sensing consists of a two-step approach to obtain the combination of speed and sensitivity. First, fast sensing is performed at speeds of approximately 1 ms per channel. Second, fine sensing is used to investigate potential empty channels with higher sensitivity and takes approximately 25 ms per channel. At first glance it might seem that the spectral scanning can be completed very fast, but since there are 135 potential channels to investigate and continuously monitor, the process can take quite some time. Conventional approaches [84,86,87] are based on single channel reception, which forces the detection to be sequential. However, the adoption of a broadband sampling receiver would speed up the process of spectral scanning significantly.

The receiver in [7] based on a 10 bit 2.6 GS/s interleaved SAR enables simultaneous down conversion of four channels, arbitrarily located in the 50–1000 MHz frequency range. Since there is no analog LO-carrier generation, hopping from one channel to the next is instantaneous, and very fast channel scanning with four channels in parallel is possible. However, for spectral scanning the integration of an ADC able to capture GHz spectrum with digital signal processing and memory functions offers another faster method, based on capturing the full spectrum. Instead of down converting on a channel-by-channel basis and measuring the power in each channel separately, the complete GHz-wide power spectrum is calculated from a small number of samples, directly taken from the full-rate ADC output.

Figure 10.13 shows a power spectrum (0–1.3 GHz range) of the digitized signal, using 16,384 samples. From the power spectrum it is now straightforward to detect the channels that are present. Since only a small number of ADC samples are required to calculate a high-resolution spectral plot, the total detection time for all channels can be as short as several μs.

To appreciate the quality of the conversion, Figure 10.14 shows a zoom-in on the channels around 850 MHz for a cable TV application scenario. The weak 256QAM-modulated channel with an SNR of around 30 dB in between two strong channels can be clearly identified. Note that no new data was captured to get this higher resolution image. The distance to the noise floor, visually identified, is large enough to detect

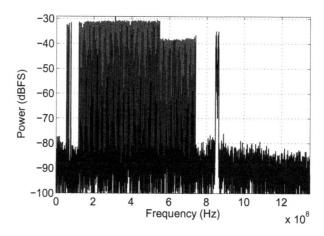

Figure 10.13

Power spectrum at the ADC output.

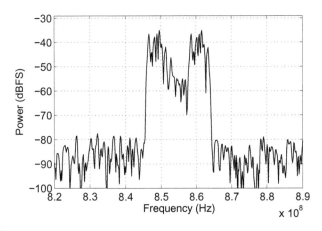

Figure 10.14

Power spectrum at the ADC output—zoomed-in to 850 MHz region.

channels with SNR values of around 0 dB. To detect weaker channels, methods with a higher sensitivity will be required. The higher sensitivity could be obtained with the same receive chain by spending more time, or alternatively by using a narrowband receiver [84]. These techniques enable today's cable TV receivers to monitor the quality of the signal and cable infrastructure, to realize fast zapping and to identify quickly the channel frequency plan, but could play a role as well in future cognitive radio receivers.

SUMMARY

In this paper we have given an overview of tradeoffs and degrees of freedom in the design of interleaved SARs in deep sub-micron CMOS technologies. We have seen that the use of many SAR units is the weakness and strength of interleaved SARs. Weakness because of matching and interconnect capacitance issues introduced, ultimately limiting the raw potential of the SAR unit especially for high-frequency multi-carrier signals. Strength because design techniques in an evolving CMOS technology offer degrees of freedom to deal with these issues, e.g. hierarchy, pipelining, multiplexing, calibrations, redundancy. These render the interleaved SAR approach a viable option for broadband conversion at high frequencies. Furthermore, we have indicated that interleaving freedom is restricted at high resolutions, meaning that the real strength of the interleaved SAR is at low to moderate bit levels with very high sampling rates. In the last two sections we presented how redundancy in the form of many SAR units offers opportunities to improve spurious performance with SAR randomization, and how broadband sampling can be applied in spectral sensing applications.

References

[1] J. McCreary and P. Gray, All-MOS charge redistribution analog-to-digital conversion techniques, *IEEE Journal of Solid-State Circuits*, vol. 10, no. 6, pp. 371–379, December 1975.

[2] W. C. Black and D. A. Hodges, Time interleaved converter arrays, *IEEE Journal of Solid-State Circuits*, vol. 15, no. 6, pp. 1022–1028, December 1980.

[3] D. Draxelmayr, A 6b 600 MHz 10 mW ADC array in digital 90 nm CMOS, in *IEEE International Solid-State Circuits Conference, Digest of Technical Papers*, pp. 264–265, 2004.

[4] K. Doris et al., A 480 mW 2.6 GS/s 10b 65 nm time-interleaved ADC with 48.5 dB SNDR up to Nyquist, in *IEEE International Solid-State Circuits Conference, Digest of Technical Papers*, pp. 180–182, 2011.

[5] P. Schvan et al., A 24 GS/s 6b ADC in 90 nm CMOS, in *IEEE International Solid-State Circuits Conference, Digest of Technical Papers*, pp. 544–545, 2008.

[6] P. Bower and I. Dedic, High speed converters and DSP for 100G and beyond, *Optical Fiber Technology*, vol. 17, no. 5, pp. 464–471, October 2011.

[7] E. Janssen et al., A direct sampling multi-channel receiver for DOCSIS 3.0 in 65 nm CMOS, in *Proceedings of the IEEE Symposium on VLSI Circuits*, pp. 292–293, 2011.

[8] N. Chawla et al., A 1 GHz digital channel multiplexer for satellite OutDoor Unit based on a 65 nm CMOS transceiver, in *IEEE International Solid-State Circuits Conference, Digest of Technical Papers*, pp. 180–182, 2009.

[9] Y. Greshishchev et al., A 40 GS/s 6b ADC in 65 nm CMOS, in *IEEE International Solid-State Circuits Conference, Digest of Technical Papers*, pp. 390–391, 2010.

[10] Y. Greshishchev, CMOS ADCs for optical communications, in *Proceedings of the 20th Workshop on Advances in Analog Circuit Design (AACD)*, April 2012.

[11] J. Stevenson et al., A multi-standard analog and digital TV tuner for cable and terrestrial applications, in *IEEE International Solid-State Circuits Conference, Digest of Technical Papers*, pp. 210–211, 2007.

[12] J. R. Tourret et al., SiP tuner with integrated LC tracking filter for both cable and terrestrial TV reception, *IEEE Journal of Solid-State Circuits*, vol. 42, no. 12, pp. 2809, December 2007.

[13] F. Gatta et al., An embedded 65 nm CMOS baseband IQ 48 MHz–1 GHz dual tuner for DOCSIS 3.0, *IEEE Journal of Solid-State Circuits*, vol. 44, no. 12, pp. 3511–3525, December 2009.

[14] F. Kuttner, A 1.2 V 10b 20 MSample/s non-binary successive approximation ADC in 0.13 μ CMOS, in *IEEE International Solid-State Circuits Conference, Digest of Technical Papers*, pp. 176–177, 2002.

[15] Z. Cao, S. Yan, and Y. Li, A 32 mW 1.25 GS/s 6b 2b/Step SAR ADC in 0.13 μm CMOS, in *IEEE International Solid-State Circuits Conference, Digest of Technical Papers*, pp. 542–543, 2008.

[16] H. Wei et al., A 0.024 mm2 8b 400 MS/s SAR ADC with 2b/Cycle and Resistive DAC in 65 nm CMOS, in *IEEE International Solid-State Circuits Conference, Digest of Technical Papers*, pp. 188–190, 2011.

[17] S. M. Louwsma et al., A 1.35 GS/s, 10b, 175 mW time-interleaved AD converter in 0.13 μm CMOS, *IEEE Journal of Solid-State Circuits*, vol. 43, no. 4, pp. 778–786, April 2008.

[18] B. Verbruggen, M. Iriguchi, and Craninckx, A 1.7 mW 11b 250 MS/s 2+ interleaved fully dynamic pipelined SAR ADC in 40 nm digital CMOS, in *IEEE International Solid-State Circuits Conference, Digest of Technical Papers*, pp. 466–469, 2012.

[19] B. Weir, Z. G. Boyacigiller, and P. Bradshaw, An error-correcting 14b/20 ps CMOS A/D converter, in *IEEE International Solid-State Circuits Conference, Digest of Technical Papers*, pp. 62–63, 1981.

[20] D. Draxelmayr, A self calibration technique for redundant A/D converters providing 16b accuracy, in *IEEE International Solid-State Circuits Conference, Digest of Technical Papers*, pp. 204–205, 1988.

[21] W. Liu et al., A 12b 22.5/45 MS/s 3.0 mW 0.059 mm^2 CMOS SAR ADC achieving over 90 dB SFDR, in *IEEE International Solid-State Circuits Conference, Digest of Technical Papers*, pp. 380–381, 2010.

[22] S.-W.M. Chen and R. W. Brodersen, A 6-bit 600-MS/s 5.3-mW asynchronous ADC in 0.13-μm CMOS, *IEEE Journal of Solid-State Circuits*, vol. 41, no. 12, pp. 731–740, December 2006.

[23] E. Alpman et al., A 1.1V 50 mW 2.5 GS/s 7b time-interleaved C-2C SAR ADC in 45 nm LP digital CMOS, in *IEEE International Solid-State Circuits Conference, Digest of Technical Papers*, pp. 76–77, 77a, 2009.

[24] B. P. Ginsburg and A. P. Chandrakasan, Highly interleaved 5-bit, 250-MSample/s, 1.2-mW ADC with redundant channels in 65-nm CMOS, *IEEE Journal of Solid-State Circuits*, vol. 43, no. 12, pp. 2641–2650, December 2008.

[25] K. Doris et al., A 480 mW 2.6 GS/s 10b time-interleaved ADC with 48.5 dB SNDR up to Nyquist in 65 nm CMOS, *IEEE Journal of Solid-State Circuits*, vol. 46, no. 12, pp. 2821–2833, December 2011.

[26] M. Harsener et al., A 14b 40 MS/s redundant SAR ADC with 480 MHz clock in 0.13 μm CMOS, in *IEEE International Solid-State Circuits Conference, Digest of Technical Papers*, pp. 248–249, 2007.

[27] C. Hurrel et al., An 18b 12.5 MHz ADC with 93 dB SNR, in *IEEE International Solid-State Circuits Conference, Digest of Technical Papers*, pp. 378–379, 2010.

[28] J. Fredenburg and M. Flynn, A 90MS/s 11 MHz bandwidth 62 dB SNDR noise-shaping SAR ADC, in *IEEE International Solid-State Circuits Conference, Digest of Technical Papers*, pp. 468–471, 2012.

[29] J. Craninckx and G. Van der Plas, A 65 J/Conversion-Step 0-to-50MS/s 0–0.7 mW 9b charge-sharing SAR ADC in 90 nm Digital CMOS, in *IEEE International Solid-State Circuits Conference, Digest of Technical Papers*, pp. 246–247, 2007.

[30] P. Harpe et al., A 30 fJ/Conversion-step 8b 0–10MS/s asynchronous SAR ADC in 90 nm CMOS, in *IEEE International Solid-State Circuits Conference, Digest of Technical Papers*, pp. 387–389, 2010.

[31] P. Harpe et al., A 7–10b 0–4MS/s flexible SAR ADC with 6.5–16 fJ/conversion-step, in *IEEE International Solid-State Circuits Conference, Digest of Technical Papers*, pp. 472–475, 2012.

[32] M. Yip and A. Chandrakasan, A resolution-reconfigurable 5–10b 0.4–1 V power scalable SAR ADC, in *IEEE International Solid-State Circuits Conference, Digest of Technical Papers*, pp. 190–191, 2011.

[33] M. van Elzakker et al., A 10-bit charge-redistribution ADC consuming 1.9 μW at 1 MS/s, *IEEE Journal of Solid-State Circuits*, vol. 45, no. 5, pp. 1007–1015, May 2010.

[34] B. Gordon, Linear electronic analog/digital conversion architectures, their origins, parameters, limitations, and applications, vol. 25, no. 7, pp. 391–418, July 1978.

[35] D. Draxelmayr, *Concepts and Improvements in Pipelined and SAR ADCs*. Kluwer Academic Publishers, 2006.

[36] A. M. A. Ali et al., A 16-bit 250-MS/s IF sampling pipelined ADC with background calibration, *IEEE Journal of Solid-State Circuits*, vol. 45, no. 12, pp. 2602–2612, December 2010.

[37] S. Bardsley et al., A 100-dB SFDR 80-MSPS 14-bit 0.35-μm BiCMOS pipeline ADC, *IEEE Journal of Solid-State Circuits*, vol. 41, no. 9, pp. 2144–2153, September 2006.

[38] N. Kurosawa et al., Explicit analysis of channel mismatch effects in time-interleaved ADC systems, *IEEE Transactions on Circuits and Systems I, Fundamental Theory and Applications*, vol. 48, no. 3, pp. 261–271, March 2001.

[39] B. Brannon and B. Schofield, Multicarrier WCDMA feasibility application note (AN-807), <http://www.analog.com>.

[40] G. Manganaro, *Advanced Data Converters*. Cambridge University Press, 2012.

[41] C. Azeredo-Leme, Clock jitter effects on sampling: a tutorial, *IEEE Circuits and Systems Magazine Third Quarter*, pp. 26–37, 2011.

[42] A. Balakrishnan, On the problem of time jitter in sampling, *IRE Transactions on Information Theory*, vol. 8, no. 3, pp. 226–236, 1962.

[43] K. Doris, *High-speed D/A Converters: From analysis and synthesis concepts to IC implementation*, PhD Thesis, Technical University Eindhoven, The Netherlands, 2004.

[44] B. Liu and T. Stanley, Error bounds for jitter sampling, vol. 10, no. 4, pp. 449–454, October 1965.

[45] N. Da Dalt et al., On the jitter requirements of the sampling clock for analog-to-digital converters, vol. 49, no. 9, pp. 1354–1360, September 2002.

[46] M. Shinagawa et al., Jitter analysis of high-speed sampling systems, vol. 25, no. 1, pp. 220–224, Febraury 1990.

[47] T. Zogakis, The effect of timing jitter on the performance of a discrete multitone system, vol. 44, no. 7, pp. 799–808, July 1996.

[48] T. M. Souders et al., The effects of timing jitter in sampling systems, vol. 39, pp. 80–85, Febraury 1990.

[49] T. Pollet et al., Effect of carrier phase jitter on single-carrier and multi-carrier qam systems, 1995.

[50] M. Bolatkale et al., A 4 GHz continuous-time $\Delta\Sigma$ ADC with 70 dB DR and 74 dBFS THD in 125 MHz BW, *IEEE Journal of Solid-State Circuits*, vol. 46, no. 12, pp. 2857–2868, December 2011.

[51] D. Stepanovic and B. Nicolic, A 2.8 GS/s 44.6 mW time-interleaved ADC achieving 50.9 dB SNDR and 3 dB effective resolution bandwidth of 1.5 GHz in 65 nm CMOS, in *Proceedings of the IEEE Symposium on VLSI Circuits*, pp. 84–85, 2011.

[52] K. Poulton et al., A 20 GS/s 8b ADC with a 1 MB memory in 0.18 μm CMOS, in *IEEE International Solid-State Circuits Conference, Digest of Technical Papers*, pp. 318–319, 2003.

[53] S. Shahramian, S. P. Voinigescu, and A. C. Carusone, A 35-GS/s, 4-bit flash ADC with active data and clock distribution trees, *IEEE Journal of Solid-State Circuits*, vol. 44, no. 6, pp. 1709–1720, June 2009.

[54] A. Nazemi et al., A 10.3 Gs/s 6 bit (5.1 ENOB at Nyquist) time-interleaved/pipelined ADC using open-loop amplifiers and digital calibration in 90 nm CMOS, in *Proceedings of the IEEE Symposium on VLSI Circuits*, pp. 18–19, 2008.

[55] C.-C. Hsu et al., An 11b 800 MS/s time-interleaved ADC with digital background calibration, in *IEEE International Solid-State Circuits Conference, Digest of Technical Papers*, pp. 464–465, 2007.

[56] C.-C. Hsu et al., A 7b 1.1 GS/s reconfigurable time-interleaved ADC in 90 nm CMOS, in *Proceedings of the IEEE Symposium on VLSI Circuits*, pp. 66–67, 2007.

[57] R. C. Taft et al., A 1.8-V 1.6-GSample/s 8-b self-calibrating folding ADC with 7.26 ENOB at Nyquist frequency, *IEEE Journal of Solid-State Circuits*, vol. 39, no. 12, pp. 2107–2115, December 2004.

[58] W. Liu et al., A 600 MS/s 30 mW 0.13 µm CMOS ADC array achieving over 60 dB SFDR with adaptive digital equalization, in *IEEE International Solid-State Circuits Conference, Digest of Technical Papers*, pp. 82–83, 2009.

[59] X. Jiang and M.-C.F. Chang, A 1-GHz signal bandwidth 6-bit CMOS ADC with power-efficient averaging, *IEEE Journal of Solid-State Circuits*, vol. 40, no. 2, pp. 532–535, Febraury 2005.

[60] Z. Cao, S. Yan, and Y. Li, A 32 mW 1.25 GS/s 6b 2b/step SAR ADC in 0.13 µm CMOS, *IEEE Journal of Solid-State Circuits*, vol. 44, no. 3, pp. 862–873, March 2009.

[61] J. Brunsilius et al., A 16b 80 MS/s 100 mW 77.6 dB SNR CMOS pipeline ADC, in *IEEE International Solid-State Circuits Conference, Digest of Technical Papers*, pp. 186–187, 2011.

[62] B. Razavi, *Principles of Data Conversion System Design*, vol. 126, New York: IEEE Press, 1995.

[63] P. Harpe, *Concepts for Smart AD and DA Converters*, PhD Thesis, Technical University Eindhoven, The Netherlands, 2010.

[64] L. Sumanen, M. Waltari, and K. Halonen, Optimizing the number of parallel channels and the stage resolution in time-interleaved pipeline A/D converters, in *IEEE International Symposium on Circuits and Systems, ISCAS*, pp. 613–616, 2000.

[65] M. Gustavsson and N. N. Tan, A global passive sampling technique for high-speed switched-capacitor time-interleaved ADCs, *IEEE Transactions on Circuits Systems II, Analog Digital Signal Processing*, vol. 47, no. 9, pp. 821–831, September 2000.

[66] S. Gupta et al., A 1 GS/s 11b time-interleaved ADC with 55-dB SNDR, 250 mW power realized by a high bandwidth scalable time-interleaved architecture, *IEEE Journal of Solid-State Circuits*, vol. 41, pp. 2650–2657, December 2006.

[67] B. Verbruggen et al., A 2.6 mW 6 bit 2.2 GS/s fully dynamic pipeline ADC in 40 nm digital CMOS, *IEEE Journal of Solid-State Circuits*, vol. 45, no. 45, pp. 2080–2090, October 2010.

[68] B. Murmann, ADC performance survey, <http://www.stanford.edu/murmann/adcsurvey.html>.

[69] C. Vogel, The impact of combined channel mismatch effects in time-interleaved ADCs, *IEEE Transactions on Instrumentation and Measurement*, vol. 54, no. 1, pp. 415–427, 2005.

[70] T.-H. Tsai, P. Hurst, and S. Lewis, Bandwidth mismatch and its correction in timeinterleaved analog-to-digital converters, *IEEE Transactions on Circuits and Systems II, Analog Digital Signal Processing*, vol. 53, no. 10, pp. 1133–1137, 2006.

[71] M. El-Chammas and B. Murmann, General analysis on the impact of phase-skew in time-interleaved ADCs, in *IEEE International Symposium on Circuits and Systems, ISCAS*, pp. 17–20, 2008.

[72] C. Enz and G. Temes, Circuit techniques for reducing the effects of op-amp imperfections: autozeroing, correlated double sampling, and chopper stabilization, *IEEE Journal of Solid-State Circuits*, vol. 84, no. 11, pp. 1584–1614, November 1996.

[73] C.-Y. Wang and J.-T. Wu, A background timing-skew calibration technique for time-interleaved analog-to-digital converters, *IEEE Transactions on Circuits and Systems II*, vol. 53, no. 4, pp. 299–303, 2006.

[74] C.-C. Huang, C.-Y. Wang, and J.-T. Wu, A CMOS 6-Bit 16-GS/s time-interleaved ADC using digital background calibration techniques, *IEEE Journal of Solid-State Circuits*, vol. 46, no. 4, pp. 848–858, April 1993.

[75] R. Payne et al., A 12b 1 GS/s SiGe BiCMOS two-way time-interleaved pipeline ADC, in *IEEE International Solid-State Circuits Conference, Digest of Technical Papers*, pp. 182–184, 2011.

[76] J. Elbornsson, F. Gustafsson, and J.-E.Eklund, Blind adaptive equalization of mismatch errors in a time-interleaved A/D converter system, *IEEE Transactions on Circuits and Systems I*, vol. 51, no. 1, pp. 151–158, September 2004.

[77] C. Vogel, V. Pammer, and G. Kubin, A novel channel randomization method for time-interleaved ADCs, in *IEEE International Symposium on Circuits and Systems ISCAS*, vol. 1. IEEE, pp. 150–155, 2005.

[78] C. Vogel and G. Kubin, Analysis and compensation of nonlinearity mismatches in time-interleaved ADC arrays, in *IEEE International Symposium on Circuits and Systems, ISCAS*, vol. 1. IEEE, pp. I–593, 2004.

[79] C. Vogel, D. Draxelmayr, and G. Kubin, Spectral shaping of timing mismatches in time-interleaved analog-to-digital converters, in *IEEE International Symposium on Circuits and Systems, ISCAS*. IEEE, pp. 1394–1397, 2005.

[80] H. Jin, E. Lee, and M. Hassoun, Time-interleaved A/D converter with channel randomization, in *Proceedings of 1997 IEEE International Symposium on Circuits and Systems, 1997. ISCAS '97*, vol. 1. pp. 425–428, 1997.

[81] K. El-Sankary, A. Assi, and M. Sawan, New sampling method to improve the SFDR of time-interleaved ADCs, in *IEEE International Symposium on Circuits and Systems, ISCAS*, vol. 1, pp. I–833, 2003.

[82] M. Tamba et al., A method to improve SFDR with random interleaved sampling method, in *International Proceedings on Test Conference, 2001*, pp. 512–520, 2001.

[83] J. Elbornsson, F. Gustafsson, and J. Eklund, Analysis of mismatch noise in randomly interleaved ADC system, in *Proceedings of the International Conference on Acoustics, Speech, and Signal Processing, ICASSP'03, 2003*, vol. 6, pp. VI–277, 2003.

[84] B. Razavi, Cognitive radio design challenges and techniques, in *IEEE Journal of Solid-State Circuits*, vol. 45, no. 8, pp. 1542–1553, 2010.

[85] IEEE 802.22 Working Group on Wireless Regional Area Networks, <http://www.ieee802.org/22/urlhttp://www.ieee802.org/22/>.

[86] J. Park et al., A fully integrated UHF-band CMOS receiver with multi-resolution spectrum sensing (MRSS) functionality for IEEE 802.22 cognitive radio applications, in *IEEE Journal of Solid-State Circuits*, vol. 44, no. 1, pp. 258–268, 2009.

[87] M. Kitsunezuka, A 30 MHz–2.4 GHz CMOS receiver with integrated RF filter and dynamic-range-scalable energy detector for cognitive radio, in *IEEE Radio Frequency Integrated Circuits Symposium (RFIC), 2011*, pp. 1–4, 2011.

High-Performance Digital-to-Analog Converters for Wireless Infrastructure

Gil Engel and Gabriele Manganaro

Analog Devices, Inc., Wilmington, MA 01887, USA

INTRODUCTION

The rapid expansion of consumer demand for data services of all types has caused service providers in wired and wireless communications to improve the data-handling capability of their networks. Cable service providers work to improve video quality from analog to digital to high definition, and then to include an internet service at higher and higher data rates, both downstream and upstream. Wireless service providers moved from analog to digital cellular to support more voice services, and now are upgrading networks to 4G and beyond to satisfy their customers' increasing demand for broadband data services on their smart phones. Backhaul service providers must upgrade their systems as the volume of data transferred between various networks increases by orders of magnitude. The common issue facing all of these markets is the consumer's expectation that these data services are available for a nominal price that is flat, regardless of the amount of data passed through the networks. Thus, cost of the systems and their upgrades is a significant factor in choosing the architecture for the radio transmission subsystem.

11.1 Transmitter architectures

While there are apparent topological similarities between the transmitter architectures covered in this section and the corresponding receiver architectures, there are also fundamental differences that need to be kept in mind. To begin with, in receiver architectures the desired signal power is generally unknown and can be very small in power. Furthermore, the desired signal may possibly be adjacent in frequency allocation to undesired signals, known as "blockers," due to other communication signals. The blockers need to be separated from the desired content (for either reception or removal) and can negatively impact the quality of the desired signal itself through, for example, intermodulation. On the other hand, in transmitter architectures, the desired signal is directly synthesized by the DAC itself and generally has the largest power in the generated transmit spectrum. Undesired or spurious content in the transmit output is either the result of the nonlinearity of the blocks constituting the transmit chain (often dominated by the power amplifier) or spurious and introduced by coupling from other sources. These fundamental differences between receive and transmit paths need to be kept in mind as the associated trade-offs and architectural choices are respectively different.

The infrastructure transmission systems have kept pace with the move to higher data rates that drive broader bandwidth and higher-order modulation, originally employing traditional heterodyne or super-heterodyne up-conversion architectures, but moving to more modern implementations. In the heterodyne-type architectures, the digital modulators are often implemented as traditional quadrature modulators, as shown in Figure 11.1a, in which a pair of in-phase and quadrature DACs drive baseband filtered data into a quadrature modulator, and the modulator's output is then up converted by one or two stages of RF mixers to the final output frequency. This signal is then amplified by a power amplifier and routed to the antenna or cable plant.

The same basic transmitter can be used to implement higher performance systems. For example, in Figure 11.1b, instead of sending baseband filtered data, the digital ASIC or FPGA can send a modulated signal and its complex conjugate to the I and Q DACs, respectively, and implement a single-sideband (SSB) upconverter, reducing or eliminating the unwanted image at the output of the upconverting mixer. This complex IF (CIF) architecture simplifies the filtering requirements and thus enables lower cost filters to be implemented. An alternate implementation keeps the signals at baseband and adds digital pre-distortion (DPD) for the output power amplifier, producing a type of zero-IF (ZIF) transmitter, though often the performance requirements of the system require the modulator to output only at IF and then be followed by an upconversion stage to reach the desired RF channel.

Any of these architectures suffers from several performance degradations that must be overcome to meet system requirements. For example, the DACs and modulators must have low enough output noise to meet system noise floor specifications. They must also have good enough balance and offset matching to avoid local oscillator (LO) leakage at the modulator output caused by amplitude or phase imbalance. The DACs often have offset and phase tuning capabilities to reduce or eliminate such imbalances. Inherent to the heterodyne architecture are the unwanted mixing products, or images, created by the upconverting mixer. These must be filtered prior to the power amplifier because they would violate in-band or out-of-band spurious suppression requirements. The DACs also have spurious signals that they produce, such as those related to the data interface clock or the DAC sampling clock. These signals will be modulated onto the carrier by the quadrature modulator, unless they are filtered well enough at the DAC outputs.

An alternate solution is shown in Figure 11.1c. In this implementation, the desired signal is created entirely in the digital ASIC or FPGA, and then an RF DAC is used to synthesize that signal directly at the desired RF output. The signal is filtered to clean up the spectrum and then sent to the transmit amplifier. The advantage of such an approach is immediately evident in its simplicity—the concept of LO leakage is eliminated, as is the upconverter's image. There is no need to balance the amplitude of a modulator's two inputs, and the phase imbalance of the quadrature modulator

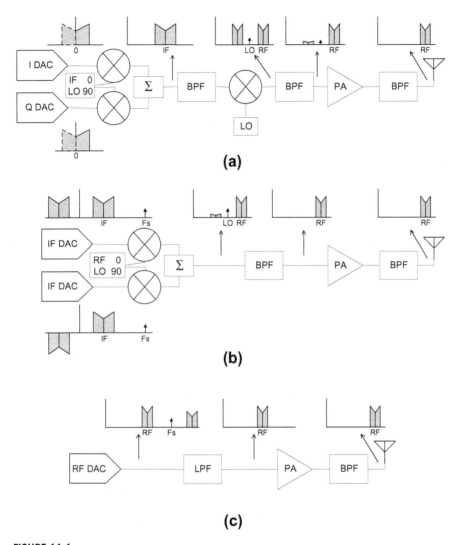

FIGURE 11.1

Radio transmitter block diagrams. (a) Traditional superheterodyne implemented with base-band DACs, (b) complex IF modulator with IF DACs, (c) direct RF synthesis with RF DAC.

similarly is no longer a concern. The modulator can be considered ideal to the level of quantization available in the ASIC or FPGA and the DAC.

The system requirements remain the same, and so the RF DAC must bear almost the entire burden of noise performance and spurious performance for the transmitter to meet the specifications. The system analysis moves from frequency planning of LO's and mixing products to planning of the DAC's noise and spurious performance and its own images and clock frequencies. Achieving these specifications while maintaining a relatively low power consumption is a key challenge. While not low power with respect to handsets, infrastructure equipment power consumption is increasingly important as service providers look to the total cost of operation.

Table 11.1 summarizes some of the most difficult specifications in cable [1] and wireless infrastructure [2] systems that drive the RF DAC requirements. For existing wireless infrastructure systems, the noise floor requirements in particular are

Table 11.1 Key Specifications in Cable and Wireless Infrastructure Standards that Drive RF DAC Performance Requirements

System	Parameter	Specification	Impact on RF DAC
Cable infrastructure	Single 256-QAM carrier in-band spurious	−73 dBc	Drives high SFDR performance
	750 kHz first ACP	−58 dBc	Has impact on minimum DAC resolution
	Single 256-QAM carrier output power	+60 dBmV (+13 dBm)	Higher DAC output power keeps driver amp gain to <30 dB
Wireless infrastructure (WCDMA)	Adjacent channel leakage ratio (ACLR) (first/second channel)	45 dB/50 dB	Drives minimum DAC resolution
	Spectral emissions mask	−56 dBc/1 MHz	Better noise floor
	Out-of-band spurious emissions	−36 dBm/MHz below 1 GHz	Higher DAC SFDR far away from desired Fout
	Noise in receive band	−98 dBm	Superior noise performance needed to reduce filtering requirements
Wireless infrastructure (MC-GSM)	Spectrum due to modulation and wideband noise	−80 dBc @ >6 MHz offset	Higher DAC SFDR and superior nearby noise performance
	GSM output power	+43 dBm	Higher DAC output power keeps PA gain <40 dB
	Out-of-band spurious emissions	−36 dBm below 1 GHz	Higher DAC SFDR far away from desired Fout
	Noise in receive band	−98 dBm	Superior noise performance needed to reduce filtering requirements

met with a combination of DAC performance and external filtering, but the DAC performance must be good enough so as not to require excessive filtering which could drive up the cost of the base stations. The wide synthesizable bandwidth requirement of cable systems creates some difficulty in DAC output design, but it also creates a high performance requirement for the DAC data interface design since the data throughput must exceed 2 GS/s to achieve 1 GHz of synthesizable bandwidth.

Practical implementations of cable or wireless infrastructure transmitters with RF DACs have been difficult to reach due to both power considerations and raw DAC performance. However, recent advances in low-power RF DAC design have made these implementations possible. Starting with narrowband implementations in cable head-ends, equipment is now deployed with the architecture of Figure 11.1c that synthesizes 4, 8, and 16 channels per DAC. With new FPGAs be coming available, the data processing necessary for a full band data throughput is now feasible from economic and power consumption view points. Thus, the focus is on the DAC design to deliver the performance summarized in Table 11.1.

11.2 **RF DAC design challenges**

The applications described above impose a difficult challenge on the RF DAC design. Some of the broadband applications require not only wide bandwidth synthesis but also demand high spectral purity. Previous applications afforded the system architect the ability to "frequency plan" by calculating where the folded back second and third harmonics would fall and adjust the sample rate such that distortions and other spurs were out of band. The resulting DAC output would then be band-pass filtered to extract the desired signal. However, requirements to output 1 GHz of bandwidth while maintaining significant out-of-band rejection near the passband provide no opportunity to move spurs and harmonics to unused portions of the spectrum and effectively hide them. In addition to the broadband signal requirements the RF DAC also targets high-frequency synthesis. These design requirements are further complicated by the need to minimize power for portable electronic devices as well as high-density communication circuit board solutions. While an in-depth description of high-speed DAC design can be found in [5], this section will attempt to provide an introductory understanding of the key issues and challenges.

11.2.1 **Principle of operation**

The current steering output structure is the prevailing choice for modern-day high-speed DACs in fine-line CMOS processes. The basic idea behind this architecture consists of synthesizing an (output) current with an amplitude that is controllable by an (input) digital word. The current generated by the DAC is used to drive follow-on RF output circuitry. In most cases a reconstruction filter directly follows the DAC aimed at suppressing images and other undesired spectral content while converting the output current to a voltage.

The preference for current steering outputs in high-speed DACs is driven by two fundamental principles: first, the output is obtained by properly summing weighted electrical quantities as directed by the input digital word, and second, adding (weighted) currents at very high speed is accomplished by shunting them all to a common node, the output.

Conversely, switch capacitor DACs commonly found in the core of most ADCs require active circuitry (e.g. operational amplifiers and operational transconductors) to realize "virtual ground" nodes that sum the weighted electric charge as well as drive subsequent stages with their output voltage. The latter approach develops accurate settled output voltages and, often, more consistent dynamic performance over the entire operating frequency range. However, this architecture is considerably more limited in speed by the lower-frequency closed-loop performance of the active circuitry, and, often occupies more active area and requires much more quiescent power for a comparable sample rate and nominal dynamic range [14]. Furthermore, significant amplifier design challenges ensue if the switched capacitor DAC's load is a low-value resistor, such as a nominal 50 or 75 Ω load found in most RF chains.

A high-level diagram depicting the main blocks constituting a modern current steering DAC is shown in Figure 11.2a. It includes an array of current sources, differential output switches, switch driving logic, clock receiver, and supporting circuitry. An array of current sources feed the differential output switches to direct current to either the positive or negative outputs. The differential output switches are shorted together at the output developing the desired total output current, namely the desired analog converted output. A data word input indicates the desired bit-by-bit current and activates the appropriate switches through the switch driving logic. The clock receiver and clock path provide the sampling edges that update the DAC output.

In order to synthesize the desired output by the addition of currents, the current sources can be designed to be identical to each other. In this case we refer to them as "unary" currents and the conversion of a digital input word will consist of steering as many of these unit currents toward the positive output node as the input digital word dictates, while all the remaining currents are steered toward the negative output. This is accomplished by converting the digital input into a "thermometric" code, namely one that has as many 1s as the numerical value that the input represents. The thermometric code directly drives the steering switches to produce the desired addition of unit currents. Clearly, an increasing digital input will result in a corresponding addition of finite unit currents and in a steadily increasing output. That is true regardless of possible individual deviations from the ideal nominal value of the individual unit currents (e.g. mismatches from one unit current to another within the array) and hence insures a "monotonic" output (the output won't decrease as a result of an increase of input and vice versa). Such a unary array of sources and switches, on the other hand, can quickly become overly complex as the number of elements grows exponentially with the DAC's resolution.

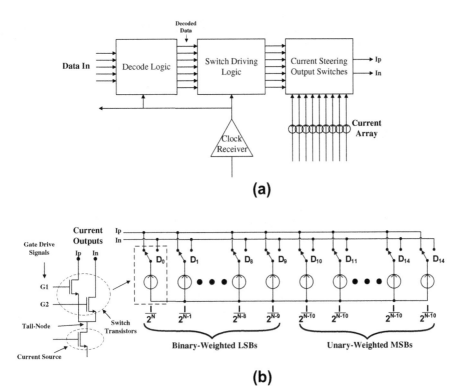

FIGURE 11.2

Top-level data flow block diagram (a), current array and output stage (b).

Another approach consists of designing a current source array for which the currents are binary weighted: namely the smallest source delivers a unit current, the following one delivers a current that is twice the intensity of the unit current, the following source is sized to deliver four times the unit current, and so on and so forth in a power of two series. In this case, each bit of the binary input code can be made to directly control the differential steering switch pair corresponding to the like weighted current source. Namely, the least significant bit (LSB) will determine the direction of the unit current source, the LSB + 1 will control the direction of the current that is twice the intensity of the previous unit, and so on, until the MSB controls the direction of the largest binary weighted source. Each weighted source directly represents one input bit and they are all summed up at the proper output node depending on the state of the input bit. While the implementation is straightforward in principle, the sizing of the currents is critical to the accuracy of the final output. For example, a 5% error in the intensity of the MSB current for a mere 8 bit DAC is far greater than the nominal size of the LSB current source (which is sized to be nominally equal to MSB/128, or ~0.8% of the MSB itself). It can be clearly apparent that a weighted array is not necessarily monotonic.

DACs used in communications range from 12 to 16 bits. Implementing (or driving) 4–65 K separate current sources is impractical, therefore DACs are typically segmented into a combination of unary and binary scaled currents. A typical current steering DAC output stage is shown in Figure 11.3. A decoder provides the appropriate mapping from the input binary word to the designed segmentation.

The performance of the DAC is primarily determined by the matching of the individual currents, their output impedance, clock noise, and timing skew. The ideal DAC will have a transfer function given by $A_0 = I \cdot Di/2^N$, errors in the output currents will deviate from the straight line transfer function causing linearity errors. Large enough deviations from the ideal transfer function will produce distortions at the output of the DAC [3–5].

Deviations from the nominal values of the current sources in the array result into errors on the settled DAC output current. Such deviations can result from manufacturing mismatches among the devices used to implement the current sources. Barring systematic design errors, these mismatches are random and the corresponding standard deviation decreases as the area of the devices increases. Device mismatch is also inversely proportional to the biasing (primarily the overdrive voltage). However, larger physical size results in larger parasitics while higher overdrive will be limited by power supply headroom. These design criteria lead to a trade-off between the accuracy of the settled output current (DC or low-frequency linearity) and the accuracy of the high-frequency (dynamic) output current, since the latter can be detrimentally affected by the magnitude of the parasitics. Alternatively, the size of the current source devices can be minimized (allowing higher manufacturing mismatches) by incorporating various types of calibration to adjust the value of each current source.

Another design consideration is the DAC output impedance, which behaves like a current divider between the external load and the DAC itself. The issue isn't the value of the impedance per se, but rather that the output impedance of the DAC is data dependent (it changes depending on the input code to the converter) and will therefore produce harmonic distortion at the output [5–8]. To minimize this effect

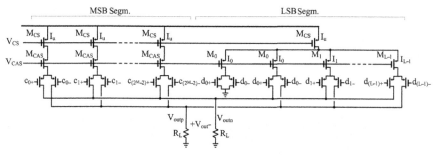

FIGURE 11.3

Simplified analog core of a high-speed DAC.

the DAC must be designed to either have minimally varying output impedance or otherwise with high enough impedance to meet the worst-case desired harmonic distortion. Since the impedance is dominated by parasitic capacitive components, it typically decreases with increasing frequency leading to worse distortion for higher frequency outputs. Hence the need, pointed out above, to keep all stray parasitics to a minimum in order to sustain linearity at high frequency.

Jitter or phase noise on the clock is yet another performance-limiting concern, as this will produce phase noise on the synthesized output signal. Additionally, any signals that corrupt the clock and/or clock path will modulate with the output and produce unwanted sidebands [9,10].

Finally, timing skew across the bits can be considered a deviation from the ideal switching instant, and will also produce distortion at the output of the DAC. Unlike current mismatch, timing skew becomes a more substantial percentage of the period as frequency and sample rate increase [11].

Timing skews can result from uneven propagation delays due to the physical location of the devices on the die or their interconnections. Timing skews can also originate from process mismatch and varying transition times of switching circuits resulting from code-dependent or mismatched loading. These can occur on the output differential switches, on the circuitry driving these switches, or at the interconnections between them. Such issues are mitigated through careful layout and signal routing as well as by minimizing timing sensitivity of the critical circuitry.

Though linearity and mismatch requirements must be taken into account, at very high frequencies demands on the clock, clock path, jitter, and other timing errors are far more challenging. The RF DAC performance is mostly limited by the dynamic errors [12].

Furthermore, large external voltage swings can couple into the DAC core and modulate with the desired output signal to produce distortion. Moreover, the output switch transistors provide a coupling path into the DAC and onto the tail node (Figure 11.2b). Any unsettled signals on the tail node will modulate with the output signal and cause distortion. The impedance and coupling effects are minimized by optimized transistor sizing and careful attention to layout.

Care must also be given to the gate drive signals. The output switch transistors route current to either the positive or negative outputs depending on whether the data is high or low. During a transition from high to low, a transient glitch will appear on the tail node due to the difference between the turn on and turn off dynamics of the output switch transistors. At very high frequency this glitch will settle differently depending on the data toggle rate. This will cause a data-dependent error and, consequently, a data-dependent distortion on the output (Figure 11.4a). This effect can be attenuated with the use of constant activity output circuitry (Figure 11.4b).

One technique for mitigating this effect is to use a return-to-zero (RZ) output circuit. Using an RZ circuit, during the first half of the period current is routed to either the positive or negative outputs depending on whether the input is high or low, and during the second half of the period the current is shunted to the supply, essentially zeroing the output. This scheme provides the constant toggling desired regardless of

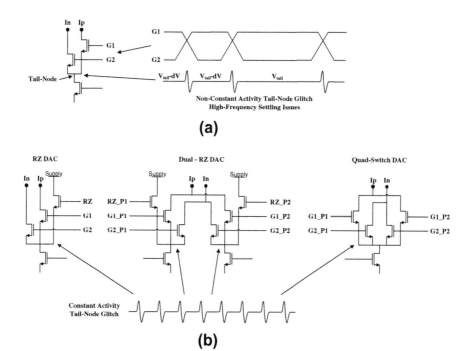

FIGURE 11.4

(a) Conventional dual switch will exhibit data-dependent tail node glitches, (b) constant activity architectures (RZ, dual-RZ, quad-switch) mitigate the data-dependent issues.

data but at the expense of throwing away half the current during the second half of the period and thus losing 3 dB of signal power at the output.

As an alternative, a dual-RZ scheme can be used in which two RZ DACs are connected in parallel, each operating at the opposite clock phase [13]. With this alternative the output power is recovered but at the expense of twice the power consumption through the output circuitry.

A third alternative uses a quad-switch output circuit, through which current switches alternatively between two pairs of output transistors. The quad-switch offers the constant switching activity that reduces data dependency without using two outputs and twice the current, making it a desirable approach.

11.2.2 Data capture and pre-processing

Design challenges for an RF DAC are not limited to the DAC output spectral performance. Applications that require large BW also require very high data throughput into the DAC interface. The data interface must be designed with length-matched and impedance-controlled data lines. In addition, the driving logic must be capable of meeting the maximum skew across the bus. For lower bandwidth applications, logic in the DAC can be used to interpolate the data to high frequency and digitally modulate the signal to the desired output frequency. In both cases the high sample rate imposes a very difficult digital-to-analog interface problem within the DAC. Care must be taken to isolate the noisy digital signals from the sensitive analog circuitry while properly sampling and retiming the digital data to the final analog output drive circuitry.

One practical implementation includes a dual port 14-bit LVDS interface that operates at 1.5 GSPS per port to provide up to 3 GSPS throughput into the DAC [12]. An input DLL locks onto the incoming data clock to track system drift and retime the data into the DAC. The interpolation included in this design provides a method of reducing the $\sin(x)/x$ attenuation at outputs close to Nyquist, while a clock mixing technique (referred to as mix-mode™) is used to modulate the incoming data to the second and third Nyquist zones. For both the interpolated and mix-mode operations the data between the digital and analog sections is transferred at up to 6 GSPS. Digital synthesized logic cannot operate at such a high speed and is designed as multiple parallel paths. Custom logic is used to multiplex the parallel paths and stream the data at the ultimate DAC rate. To maintain a lock across frequency, process, voltage, and temperature variations, a redundant constantly toggling bit is added to the interface between the digital and analog domains (Figure 11.5). This constant-toggle bit is phase compared against the analog clock by a phase detector. The output of the phase detector imparts a control voltage to a VCO on the digital side to push or pull the frequency as needed and maintain lock between the digital and analog circuitry.

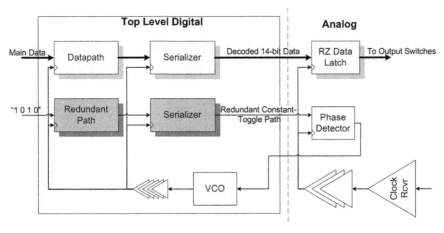

FIGURE 11.5

Data interface design is critical to high performance in RF DACs.

11.3 **Package**

To maintain RF DAC performance, the package needs to be optimized in conjunction with the silicon. In many cases, trade-offs are made to both the silicon and package to provide the best overall solution. For example, it is desirable to have the clock inputs as close to the clock receiver circuitry as possible, and simultaneously place the clock receiver circuitry as close to its final destination as is practical. However, these desires would make external package as well as board layout issues more challenging and costly, and they could potentially degrade isolation. Similar concerns apply to the DAC outputs and the power domains.

To address these trade-offs, it is desirable to choose a package that can be customized while simultaneously providing good signal integrity. This provides flexibility with signal delivery, controlled impedance where necessary, and low impedance for the power supplies. The power supply domains need to be designed and routed such that current loops are low impedance and do not couple onto other domains or signals. This requirement is ideally extended to the PCB, such that the system designer has an easier time routing power to the device and providing decoupling.

The DAC outputs must be handled very carefully. The optimal output design would use a controlled impedance in the package and take into consideration the entire signal path from the die through the package and package-to-board bond path, and finally the board design itself. Three-dimensional simulation and analysis software can be used to ensure proper transmission line design for these signals. A similar design methodology is used for the clock inputs. The S-parameter data for these signals is optimized to maximize the delivered signal power and minimize reflections.

11.4 **Practical design considerations**

Typical communication systems include a data backbone to provide data communication to a switch fabric at the top level, with data rates reaching multi-terabit data transfers. Data is routed through the appropriate channel cards and ultimately to the desired RF port. Within a channel card, the RF DAC acts as the interface between the digital logic and the RF analog output driving network. These communication system cards typically include data interface logic, a field programmable gate array (FPGA) or specialized ASIC, DACs, filters, gain blocks, and RF power amplifiers. To optimize system performance, analysis now includes the system printed circuit board and signal chain components. Models are used to simulate the driver and receiver characteristics, while 3-D simulation and analysis software is used to ensure proper transmission line design and a good return path.

High-speed data from the digital interface logic drives the DAC inputs. The digital data path is differential to maximize throughput by ensuring minimum charge transfer between the driving logic and the DAC inputs, thereby reducing distortion of the input data signals. The data path between the interface logic and the DAC is impedance-controlled and data line lengths need to be matched to minimize skew at high frequency.

Power supply routing can be extremely challenging. The digital logic includes I/O and core logic power supplies while the RF output network can include as much as four or five additional power supplies. The power domains must be isolated from one another and signal return paths need to be carefully managed to ensure no crosstalk between domains. A working knowledge of these system requirements helps guide port and supply designs in the RF DAC to ease integration.

The main DAC clock is among the critical signals on the system card. The DAC clock is differential and is isolated from other signals with via fences and controlled return paths to ensure no coupling or crosstalk. Any signals coupling onto the clock will directly appear at the output of the DAC. Digital signals corrupting the clock reduce the noise margin in the system. Even the DAC outputs must be prevented from coupling onto the clock as this will cause second harmonics and potentially other harmonic problems on the output spectrum. It is preferable to keep the clock driver as close as possible to the DAC to reduce noise and other coupling concerns.

The DAC outputs act as transmission lines between the DAC and its primary load. Close attention is paid to the DAC output network transmission lines to maintain constant impedance. As stated above, the RF DACs extend this analysis through the laminate onto the board to provide a minimal impedance difference out of the device. Matched impedance between the DAC and the load is critical in order to maximize the power transfer from the DAC to the destination and to minimize reflection from the destination back to the DAC. If the DAC and load are designed to meet $50\,\Omega$ impedance, the transmission line must be matched and designed for $50\,\Omega$ as well, such that $Z_S = Z_L = Z_{line}$ (Z_s—source impedance, Z_L—load impedance, Z_{line}—transmission line impedance). Transmission lines at RF frequencies must be treated as complex impedances having a resistive component (the real part) and reactive component (the imaginary part). For example if the driving source in the system is purely resistive but the load has a reactive component, the transmission lines need to be designed to compensate for the load with an opposite reactance and thereby maintain a matched impedance. The RF DAC, transmission lines, and load provide three distinct sections that can be treated as multi-port interfaces.

S-parameters provide the designer with a tool that can be used to optimize these interfaces. Three-dimensional simulation tools are used to extract and analyze the transmission line S-parameters and can also be used to include the source and load S-parameter data for full network simulation. S-parameters are also measured in RF systems with a network analyzer and then analyzed using the 3-D simulation tools. For differential input and output systems, two differential S-parameter ports or four single ports are required. S-parameters include information on the reflected incident power waves as well as the crosstalk between the lines. A Smith chart is used to provide a quick graphical representation of the complex impedance across frequency. S11 is a measure of the input complex reflection coefficient when the output port is terminated with the match system load while S22 is the output complex reflection coefficient when the input port is terminated with a match system load. S12 and S21 are forward and reverse transmission gain respectively. In a matched system the S11 and S22 parameters will be low and the S12 and S21 parameters will be close to zero (measured in dB, assumes 1 times the signal energy is transmitted). In an ideal system, the S11 and S22 parameters will be negative infinity and the S12 and S21 parameters will be equal to zero, which indicates that all the power transfer from the source to the destination is done without any loss.

Even though the DAC is commonly considered a differential circuit, the RF DAC also outputs common-mode signal components that need to be taken into account. The RF DAC has significant common-mode signal content at even harmonic multiples of the fundamental and at multiples of the DAC sample rate. External passives as well as the board parasitic matching are of significance in maintaining common mode rejection. In systems that include an amplifier or gain stage driven by the RF DAC, the common-mode rejection of these devices must be taken into account. Ideally, the differential amplifier output will cancel the common mode by taking the difference of the two inputs and amplifying them, as $V_o = \text{Gain} \cdot (V^+ - V^-)$. However, the amplifiers also contain a common-mode gain (that varies across frequency) and the output is described as $V_o = \text{Gain} \cdot (V^+ - V^-) + \frac{1}{2}A_{cm} \cdot (V^+ - V^-)$. Typically, the common-mode gain is much smaller than the amplifier gain, which thus provides improvement to the common-mode rejection of the system.

11.5 Measured results

An example of a state-of-the-art high-frequency/high-performance DAC has recently been presented in [12]. Figure 11.6 shows the output of a 3.0 GSPS RF DAC that is synthesizing 158 channels of 6 MHz wide, 256 QAM signals that are compliant with the DOCSIS specification. The signals are generated in an FPGA and then directly synthesized by the RF DAC. The slight ripple of an on-chip digital filter to suppress out-of-band signals can be seen. The DOCSIS 3 documentation specifies varying adjacent channel leakage ratio requirements (ACLR), which is a measurement of desired channel signal power to adjacent nontransmitted channels over a specified bandwidth. The ACLR requirement for 158 channels simultaneously is ~51 dBc; the RF DAC meets this requirement. A direct to RF signal is shown in Figure 11.6b with the RF DAC sampling at 2.4576 GSPS outputting 25 MHz-wide WCDMA channels at 884–894 MHz. Figure 11.6c shows four WCDMA signals being directly synthesized from an RF DAC at 1970–1990 MHz output frequency. The ACLR specifications for WCDMA are 45 dBc and 50 dBc at the antennae for the first and second adjacent channels respectively. Figures 11.6b and c shows the RF DAC clearly exceeding the specification with enough margin for the following RF amplification circuitry. SFDR and IMD for this DAC are shown below in Figure 11.7.

A die photo of the DAC is shown in Figure 11.8. This DAC is an intrinsic matching DAC, namely the matching of the devices implementing the current sources is obtained through proper sizing and biasing of such devices and does not require calibration or trimming. This array of current sources can be clearly seen in the center of the photo. These currents are fed to the array of differential switches on its right-hand side and finally are routed toward the output nodes visible as the wide metal

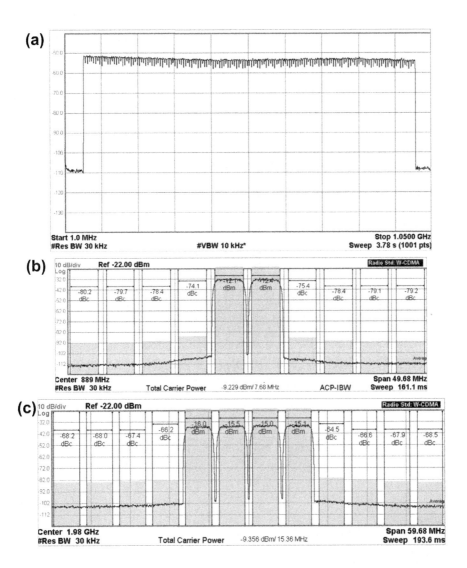

FIGURE 11.6

Measured spectrum analyzer plots of an RF DAC output. (a) 158 channels of 6 MHz, 256 QAM cable TV signals from 55 MHz to 1 GHz. (b) Two 5 MHz wide WCDMA channels at 884–894 MHz. (c) Four 5 MHz wide WCDMA signals at 1970–1990 MHz.

lines departing from the array and heading up to the output pads. The drivers for the switches, the RZ-latches, and the digital encoders controlling them are visible on the right-hand side of the switches array. The active area of this $0.18\,\mu m$ DAC is only $4\,mm^2$ and extensive description of its performance can be found in [12].

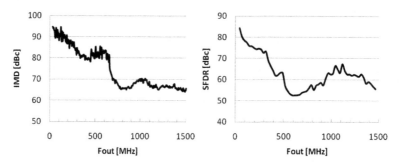

FIGURE 11.7

IMD and SFDR for a state-of-the art 3.0 GSPS RF DAC.

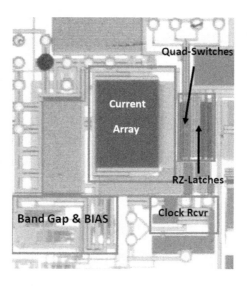

FIGURE 11.8

Die photo and block diagram of the 3.0 GSPS RF DAC.

11.6 Future direction

RF DACs are being used in today's communications infrastructure systems, as well as in wired systems such as DOCSIS cable distribution. Wireless communications systems are requiring wider data bandwidths to be implemented by service providers across more diverse frequency bands. The relatively small allocations of system bandwidths and more complicated modulation schemes are driving a push to smaller coverage areas and cell sizes while requiring the transmission systems equipment to be flexible in RF output frequencies. An RF DAC has unique advantages to address these needs. By synthesizing the entire RF output band, the RF DAC enables flexibility in modulation type, bandwidth, and output frequency. The system design becomes simpler as the modulator is implemented digitally in an FPGA or ASIC and then output through the RF DAC. The only components that change per band combination are the band-specific roofing filters and the output power amplifiers. The equipment designers can focus on optimizing those components knowing that the output signal is crafted digitally.

As for the circuit design of RF DACs, the demand for ever higher bandwidth converter with sustained dynamic performance (both linearity and noise) is without question. There is no one right answer to the many challenges we have pointed to in the above. When comparing CMOS 12–14 b DACs clocking at hundreds of megahertz designed in the late 1990s with those clocking well beyond the GSPS of today on the same CMOS process node, it is apparent that innovative circuit techniques, proper system-level partitioning and segmentation have been able to provide very significant speed and dynamic performance improvements as described above. That said, the fine lithography CMOS process offers both new circuit design challenges as well as many opportunities for further extending the high-frequency performance of RF DACs.

The main circuit-level challenges originate from some of the following limitations. First of all, the nominal power supplies for the shorter channel devices have been decreasing over subsequent CMOS process generations (as shown in Figure 11.9). On the other hand, the MOS threshold voltages have not scaled at the same rate. That has led to headroom limitations, especially when considering that the analog DAC core circuitry is constituted by a stack of MOS devices that are all meant to be operating in saturation at all times while, at the same time, leaving sufficient headroom for a large swing output (Figure 11.3). This issue leaves little headroom for cascoding. Secondly, the intrinsic gain of MOS devices, gm/gds, has dramatically dropped over subsequent process nodes to limited values on the order of 10 or less in most recent nanometer nodes. This in turn makes even a single cascode level rather less effective. To complicate matters, nanometer effects such as gate leakage or drain-induced barrier lowering make the biasing and operation of multiple sections of the DAC harder to design and to control over varying operating conditions.

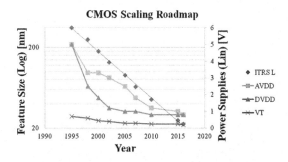

FIGURE 11.9

Critical supplies and threshold voltages for scaled CMOS processes.

There are, however, a number of important benefits brought in by scaling. The device's unity gain frequency, f_T, improves, in other words devices are faster. While the balance between devices' own parasitics and interconnect strays shifts toward an increasing relative contribution of the latter, overall, parasitic capacitances decrease together with relative distances. Also, device matching continues to improve, both in terms of mismatch due to lithography as well as threshold voltage mismatch, hence allowing better intrinsic matching for a given geometry.

Most importantly, the power efficiency of digital functionality has been increasing tremendously and the corresponding physical area has been decreasing. This benefit coincides favorably with the trends demanding more and more digital functionality and signal processing. For example, in state-of-the-art CMOS processes such as 65 nm or 40 nm it is conceivable to have a simple embedded processor in the area that would normally be required by a bonding pad. Access to this rather large computational power opens the door to many opportunities. In other words it becomes conceivable to embed digital algorithmic correction of the analog core nonlinearity together with the analog core itself at a cost, in silicon area as well as overall power consumption, that decreases over process nodes. This is part of what has been termed digital assisted analog [5,15].

References

[1] Data-Over-Cable Service Interface Specifications, Downstream RF Interface Specification, no. 12, CM-SP-DRFI-I12-111117, Cable Television Laboratories, inc., 17 November 2011.

[2] 3GPP TS 45.005 Radio Transmission and Reception (Release 10), v10.4.0, March 2012.

[3] B. Razavi, *Principles of Data Conversion System Design*. Piscataway, NJ: IEEE Press, 1995.

[4] Rudy J. van de Plassche, *CMOS Integrated Analog-to-Digital And Digital-to-Analog Converters*, 2nd ed. Dordrecht, The Netherlands: Kluwer, 2003.

[5] G. Manganaro, *Advanced Data Converters*. Cambridge University Press, 2011.

[6] A. Rodriguez-Vazquez, F. Medeiro, and E. Janssens, *CMOS Telecom Data Converters*. Kluwer Academic Publishers, 2003.

[7] C.-H. Lin, A 12-bit 2.9 GS/s DAC with IM3<−60 dBc beyond 1 GHz in 65 nm CMOS. *IEEE Journal of Solid-State Circuits*, December 2009.

[8] S. Luschas and H.-S. Lee, Output impedance requirements for DACs, in *IEEE International Symposium of Circuits and Systems*, vol. 1, pp. 861–864, 2003.

[9] G. Engel, *The Power Spectral Density of Phase Noise and Jitter: Theory, Data Analysis, and Experimental Results*, Analog Devices, AN-1067.

[10] P. Smith, *Little Known Characteristics of Phase Noise*, Analog Devices, AN-741.

[11] K. Doris, Mismatch-based timing errors in current steering DACs, in *IEEE Proceedings of International Symposium on Circuits and Systems*, 2003.

[12] G. Engel et al., A 14b 3/6 GHz current-steering RF DAC in 0.18 μm CMOS with 66 dB ACLR at 2.9 GHz, in *IEEE International Solid-State Circuits Conference*, 2012.

[13] K. Poulton, A 7.2-Gsa/s, 14-bit or 12-Gsa/s, 12-bit DAC in a 165-GHz ft BiCMOS process, in *VLSI Symposium*, 2011.

[14] G. Manganaro et al., A dual 10-b 200-MSPS pipelined D/A converter with DLL-based clock synthesizer, *IEEE Journal of Solid-State Circuits*, vol. 39, no. 11, pp. 1829–1838, November 2004.

[15] B. Murmann, A/D converter trends: power dissipation, scaling and digitally assisted architectures, in *IEEE Custom Integrated Circuits Conference*, pp. 105–112, 2008.

Time-to-Digital Conversion for Digital Frequency Synthesizers

12

Michael H. Perrott

Masdar Institute, Abu Dhabi, UAE

INTRODUCTION

Digital phase-locked loops have recently emerged as a viable alternative to traditional analog structures when implementing frequency synthesizers for wireless communication. Figure 12.1 illustrates a block diagram for a digital fractional-N frequency synthesizer. This PLL structure achieves feedback by comparing the relative time difference between edges of a reference frequency and the frequency-divided output of a digitally controlled oscillator (DCO) through the use of a time-to-digital converter (TDC). The TDC output, in turn, is passed into a digital loop filter and then into the digital tuning control of the DCO. When a constant divide value is used, the PLL structure is referred to as an integer-N PLL. By dithering the instantaneous divider value according to the output of a digital $\Delta\Sigma$ modulator, fractional divider values can be realized such that high-resolution frequency control of the synthesizer is achieved. In this case, the PLL structure is referred to as a fractional-N frequency synthesizer and will be the primary application vehicle for our TDC discussion.

A key advantage of a digital PLL implementation is that its digital loop filter, in contrast to analog loop filters used within traditional PLLs, can be realized in a small area and without concern for leakage current or noise sensitivity. However, excellent PLL phase noise performance and linear PLL behavior demand a high-resolution TDC that has well-matched quantization levels. Also, the input time range of the TDC must be adequately large to accommodate phase variations caused by dithering of the divider values and large time offsets encountered during initial settling of the PLL.

To better understand the operation and accompanying performance issues of a TDC, Figure 12.2 displays a classical delay line TDC and its DC transfer characteristic. Operation of the TDC consists of passing the edge of a $Start(t)$ signal into a chain of delay buffers whose outputs are fed into a set of registers. The edge of a $Stop(t)$ signal is used to clock the registers, which leads to an output thermometer code that is a function of the relative time difference between the $Start(t)$ and $Stop(t)$ edges. The thermometer code bits are added together to form the overall TDC output. Resolution of the time difference measurement is set by the delay buffers and is defined as T_q. Since the TDC output is integer-valued and the input resolution is T_q, the average

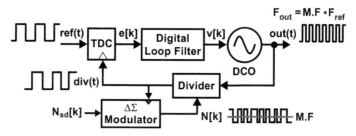

FIGURE 12.1

Digital implementation of a fractional-N frequency synthesizer PLL.

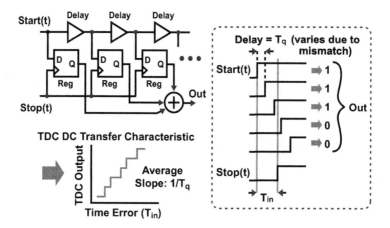

FIGURE 12.2

Classical delay line TDC.

slope of the DC transfer characteristic is $1/T_q$. However, any mismatch between the delay buffers leads to nonlinearity in the DC transfer characteristic.

The impact of TDC resolution on PLL performance can be seen by examining a linearized model of the PLL as shown in Figure 12.3. In this linearized model, the phase error between the reference and divided DCO output is converted to a time difference by the scale factor $T/(2\pi)$, where T is the period of the reference clock. TDC quantization noise, labeled as $t_q[k]$, is added, followed by the TDC scale factor of $1/T_q$. The double-sided spectrum of the TDC quantization noise, $S_{t_q}(e^{j2\pi fT})$, is computed based on a white noise assumption as

$$S_{t_q}(e^{j2\pi fT}) = \frac{1}{12}T_q^2. \tag{12.1}$$

Given the above, we see that high resolution, i.e. a low value of T_q, is highly desirable for the TDC in order to obtain low magnitude for its quantization noise spectrum.

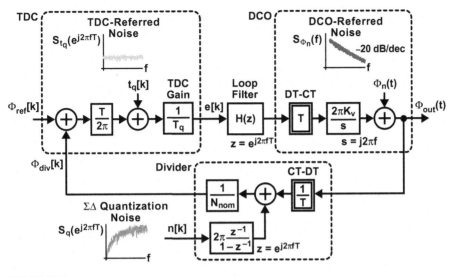

FIGURE 12.3

Linearized model of a digital fractional-*N* frequency synthesizer.

It is worth noting that TDC quantization noise has a similar impact on the phase noise of a digital PLL as charge pump noise has on the phase noise of an analog PLL.

To better clarify the impact of the various PLL noise sources on output phase noise, Figure 12.4 depicts their associated transfer function relationships, where f_o corresponds to the closed loop bandwidth of the digital PLL. We see that the TDC noise is lowpass filtered by the PLL, and that it typically sets the phase noise performance at low offset frequencies according to its resolution, T_q. As such, high TDC resolution is required to achieve low PLL phase noise at low offset frequencies. We also see that the shaped quantization noise produced by the divider dithering action of the $\Delta\Sigma$ modulator is lowpass filtered by the PLL. However, any TDC mismatch of the delay cells can lead to folding of the shaped quantization noise such that the PLL phase noise is increased at low offset frequencies. Therefore, in addition to high TDC resolution, good matching of TDC delay values is an important performance issue.

Given the above motivation, this chapter will present a gated ring oscillator (GRO) TDC as well as recent follow-on architectures which achieve the high TDC resolution and matching characteristics required for high-performance digital PLL structures such as fractional-*N* frequency synthesizers. We begin by providing background on classical TDC implementation techniques which highlight their key performance issues. The GRO TDC architecture is then presented along with discussion of its performance attributes and preferred implementation techniques. Follow-on TDC architectures are then presented as well as observations related to the commonly used Type II PLL structure, and then the chapter is concluded.

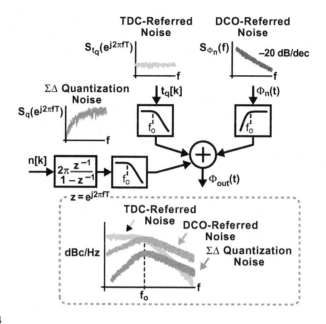

FIGURE 12.4

Transfer functions from digital PLL noise sources to output phase noise.

12.1 Background on TDC architectures

The classical delay line TDC shown in Figure 12.2 has a resolution of T_q, which is set by the delay value of its individual stages. While such delay resolutions are improving with advanced CMOS processes, alternative circuit architectures can provide a more immediate path to better performance. To better understand such architectural approaches, this section provides general background information on TDC techniques such as vernier delay cells, interpolation, time-based amplification, and oscillator-based designs.

12.1.1 Vernier technique

Figure 12.5 displays a TDC which utilizes the Vernier technique of achieving improved delay resolution by comparing the *difference* in delay between buffers. As shown in the figure, the delay comparison is performed by sending both *Start(t)* and *Stop(t)* signals into separate delay chains with individual delay values of *Delay* and *Delay2*, respectively. The change in delay seen by each register becomes *Delay–Delay2*, so that significant reduction in the effective delay per stage can be achieved.

An important implementation issue for Vernier approaches is that high accuracy is required in setting the relative delay between *Delay* and *Delay2*. A typical approach to address this issue is to employ a delay-locked loop (DLL) circuit to tune

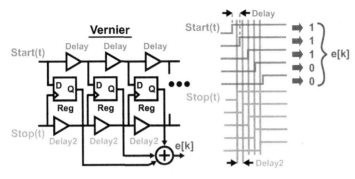

FIGURE 12.5

Vernier TDC.

each delay chain [1]. Therefore, the complexity of a vernier TDC is often significantly higher than for the classical delay chain TDC due to the need for extra calibration circuitry. Also, while the DLL circuits allow accurate setting of the *average*
value of *Delay* and *Delay2* within the two delay chains, the overall design becomes
much more sensitive to mismatch due to the reduced resolution that such mismatch is
compared to. For instance, if a vernier TDC aims to improve resolution by a factor of
four by setting $Delay = 1.25 \cdot Delay2$, then the impact of mismatch is roughly $4\sqrt{2}$
times higher in comparison to a single delay chain TDC due to the reduced resolution
and the fact that two rather than one delay chain contributes to mismatch.

As a last point, the improved resolution offered by the vernier technique also
comes at the cost of requiring a larger number of delay stages to achieve a given input
range. For instance, a factor of four improvement in resolution also leads to a factor of four increase in delay stages for the same input range, which implies a factor
of four increase in area and power as well as an increase in latency. Unless a small
input range is acceptable, practical vernier techniques must often be combined with
other circuit architecture techniques that improve the trade-off between resolution
and power/area consumption.

12.1.2 Interpolation

A different approach to improving TDC resolution is to leverage the technique of
interpolation as shown in Figure 12.6 [2]. Here a passive approach is used to combine
the outputs of a standard single delay chain TDC such that new edges are introduced
between the intrinsic edges of the delay chain. As shown in the figure, it is quite
straightforward to achieve a factor of two improvement in resolution using this technique, and even higher resolution improvement is possible.

Similar to the vernier technique, using interpolation to improve resolution while
keeping the input range constant comes at the cost of higher area and power due to
the need for additional comparison stages. Also, mismatch will have a higher impact

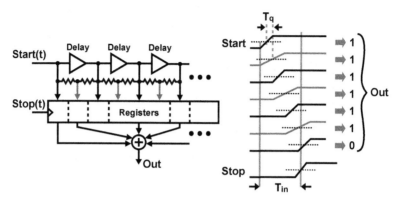

FIGURE 12.6

Delay line TDC with interpolation.

due to the reduced resolution that it is compared to. However, the interpolation technique carries the advantage that DLL calibration circuits are not required for modest improvements in resolution. Also, the impact of mismatch is somewhat reduced by the fact that the interpolation circuits tend to average the influence of mismatch between the intrinsic edges they are generated from.

12.1.3 Time-based amplification

To address the trade-off between resolution and power/area as discussed above, a two-step architecture can be employed in which coarse followed by fine measurement is performed. For instance, to implement a 10-bit design, the TDC architectures discussed so far would require $2^{10} = 1024$ comparison stages. In contrast, a two-step architecture might only need a 5 bit coarse and 5 bit fine resolution structure yielding $2 \cdot 2^5 = 64$ comparison stages. As such, considerable area and power savings are possible with this approach. However, two-step architectures require a fine resolution structure that either has significantly higher resolution than the coarse structure, or amplification of the error residue generated by the coarse structure. Recently, there has been much interest in TDC architectures that use the latter approach, though an example of the former approach is given in [3].

Figure 12.7 shows a two-step TDC architecture based on time amplification [4] of the coarse stage error residue such that the coarse and fine stages can be implemented with the same resolution [5]. To explain, the coarse measurement determines the time value within its resolution as set by a corresponding single delay chain TDC. The resulting residual error is then amplified by the *time amplifier* by a gain corresponding to the desired increase in resolution required of the fine structure. The amplified time signal is then measured by the fine structure, which also corresponds to a single delay chain TDC. A properly scaled version of the fine measurement is then added to

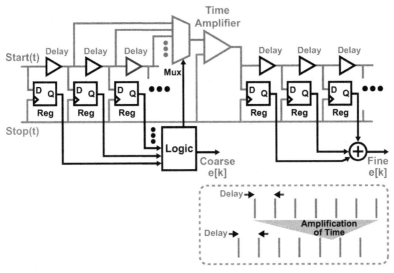

FIGURE 12.7

Time-amplifier-based TDC.

the coarse measurement to achieve the desired overall TDC output. One should note that for any two-step architecture, extensive calibration is often required in order to achieve accurate results.

Figure 12.8 displays the time amplifier proposed by [4]. This circuit leverages the metastability behavior of a common latch circuit to create an output edge whose transition time changes dramatically with the relative timing between a clock and input edge. A key issue, however, is that there is only a small time window over which such amplification occurs, which further increases the need for calibration.

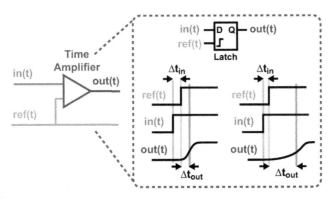

FIGURE 12.8

Time amplifier concept.

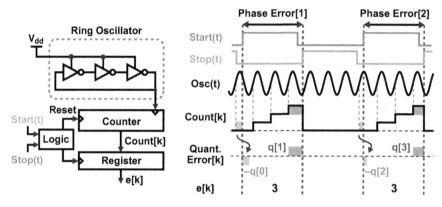

FIGURE 12.9

Oscillator-based TDC.

Overall, the time-amplifier-based two-step TDC can achieve high resolution performance [5], but challenges remain in achieving low power consumption and a low area implementation.

12.1.4 Oscillator approach

In cases where resolution is not as important as achieving low area, the oscillator-based TDC shown in Figure 12.9 offers a simple approach to achieve very wide input range with only a small number of delay stages. In this case, time is measured by counting the number of oscillator cycles that occur between the *Start*(*t*) and *Stop*(*t*) edges, so that the available input range is set by the number of bits in the counter. As an example, a 10-bit single delay chain TDC would require $2^{10} = 1024$ comparison stages, but an oscillator-based TDC would only require a few delay stages and a 10-bit counter.

While the example design shown in Figure 12.9 offers only coarse resolution corresponding to the period of the oscillator, higher resolution is readily obtained by placing rising-and falling-edge-activated counters at the output of each delay stage. In doing so, the delay resolution improves to that of a single delay stage.

An interesting characteristic of the oscillator-based TDC is that the quantization error it produces is scrambled by virtue of the varying phase of the oscillator across different measurement periods, which is illustrated in Figure 12.9. We will elaborate on this issue further in the next section.

12.2 Gated ring oscillator TDC

When considering TDC architectures for digital PLL applications, a key observation is that the PLL will filter the TDC output such that the average behavior of the TDC is more important than its "one-shot" measurement performance. As Figure 12.10

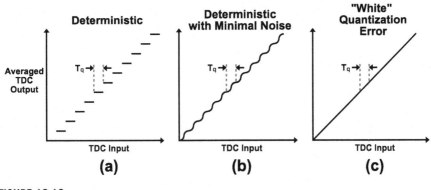

FIGURE 12.10

Comparison of TDC characteristics for (a) deterministic quantizer behavior, (b) moderate level of noise added to quantizer, (c) scrambling of quantizer error.

reveals, the filtering effect can be exploited to achieve a much more continuous TDC characteristic than the classical staircase function of a typical quantizer. To explain, part (a) of the figure illustrates a TDC characteristic with deterministic input-to-output behavior as would be encountered with the majority of TDC architectures discussed in the previous section. Part (b) of the figure shows the impact of adding a moderate amount of noise to the quantizer input. Due to averaging, the noise smooths out the TDC characteristic according to its variance versus the quantizer step size. Part (c) of the figure shows the impact of having scrambled quantization noise as would be encountered with an oscillator-based TDC. Since the oscillator phase changes across different measurements, the resulting variation of the quantization error dramatically smooths the filtered TDC characteristic such that the TDC staircase behavior is eliminated. The resulting linearity of the TDC characteristic is highly desirable as it leads to linear PLL dynamics.

This section will present a gated ring oscillator (GRO) TDC architecture that takes advantage of the scrambling property of oscillator-based TDCs and adds to it the ability to achieve first-order noise shaping of the quantization noise and delay stage mismatch. The noise-shaping property of the GRO TDC allows significant reduction of the impact of quantization noise and mismatch in digital PLL applications which filter the TDC output. In addition to presenting key concepts of the GRO TDC and preferred implementation strategies, measured results are provided from a prototype chip designed in 130 nm CMOS.

12.2.1 Key concepts

As shown in Figure 12.11, a GRO TDC is achieved by modifying an oscillator-based TDC such that its state is stopped and preserved between measurements. We refer to this action as "gating" the oscillator. Since the phase of the oscillator varies across

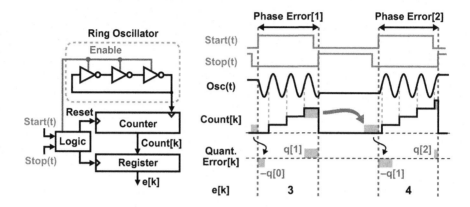

FIGURE 12.11

Gated ring oscillator (GRO) TDC.

different measurements, desired scrambling of the quantization occurs as encountered with a classical oscillator-based TDC. However, a key advantage of preserving oscillator state between measurements is that the quantization error at the end of the previous measurement is carried forward to the start of the current measurement such that the effective measurement error, $e[k]$, becomes

$$e[k] = q[k] - q[k-1], \qquad (12.2)$$

where $q[k]$ and $q[k-1]$ are defined as the quantization error at the end of the current and previous measurement, respectively.

Equation (12.2) reveals first-order noise shaping of the quantization error, which provides significant improvement in resolution when filtering is applied to the TDC output. Intuitively, first-order shaping pushes much of the quantization noise energy to high frequencies, where filtering eliminates it.

Preservation of the oscillator state is achieved with a fairly simple CMOS implementation as shown in Figure 12.12. Here a set of NMOS and PMOS switches are added to the inverters of a ring oscillator. Part (a) of the figure illustrates normal oscillator operation *within* a time measurement in which the switches are closed. Part (b) of the figure illustrates the hold state of the oscillator *between* measurements in which the switches are open and charge on the parasitic capacitors within the oscillator retains its state information. As with any dynamic CMOS circuit, leakage paths will cause the charge to gradually drain off the parasitic capacitors such that the oscillator state can become corrupted if it is left in this state for a long time. However, assuming a reasonably high-speed sample clock for the GRO TDC, the effects of leakage can be mitigated and the gating action leads to significant noise shaping benefits.

To achieve improved resolution for the GRO TDC, it is desirable to make use of all of its inverter cells when making time measurements. As pointed out in the

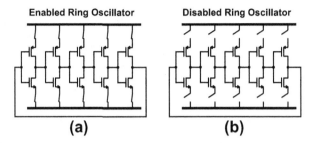

FIGURE 12.12

Implementation of GRO TDC with (a) enabled oscillator during measurements, (b) disabled oscillator between measurements.

previous discussion of oscillator-based TDCs, we simply place rising and falling edge counters at the output of each inverter as shown in Figure 12.13. As illustrated in the figure, the gating action of the oscillator still successfully leads to first-order noise shaping of the quantization noise.

An additional benefit of the GRO gating action is that it leads to first-order shaping of mismatch between the delay stages. To explain, holding the oscillator state *between* measurements causes the progression of transitions through the oscillator *during* measurements to become barrel shifted as illustrated in Figure 12.14. It is well known that barrel shifting through elements causes first-order shaping of their mismatch [6]. As such, a GRO TDC offers excellent immunity to delay mismatch

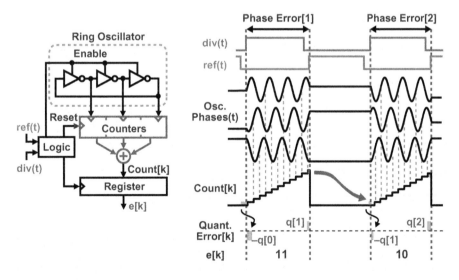

FIGURE 12.13

GRO TDC utilizing all delay stages.

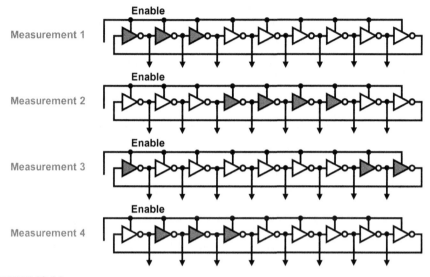

FIGURE 12.14

Barrel shifting of delay cells for GRO TDC.

since the impact of such a mismatch is pushed to higher frequencies, where it is filtered by the PLL.

12.2.2 High-performance implementation techniques

While noise shaping of the TDC quantization error substantially improves the effective TDC resolution after filtering, further improvements in resolution can be obtained by reducing the delay per inverter stage. While advances in CMOS processes offer reduced gate delay, it is worthwhile to consider improved circuit architectures to achieve better resolutions.

Figure 12.15 compares a classical ring oscillator structure to one that utilizes a multi-path topology. The key idea behind the multi-path topology is that one can use the transition information of previous delay stages to predict when a current delay stage should begin its transition and thereby reduce the effective delay time through each delay stage. Using this technique, a significant increase in operating frequency has been demonstrated for ring oscillators [7,8], and a multi-path GRO implementation has been presented which achieves roughly a five times reduction of effective gate delay [9].

Reduction of gate delay time improves TDC resolution, but carries the cost of higher operating frequencies and therefore higher power. Utilizing counters at each delay stage as shown in Figure 12.13 is particularly problematic for power dissipation when such high operating frequencies are involved. Given this issue, it is worthwhile considering alternative methods of recording transition information within the oscillator such as the phase sampling method shown in Figure 12.16. In this

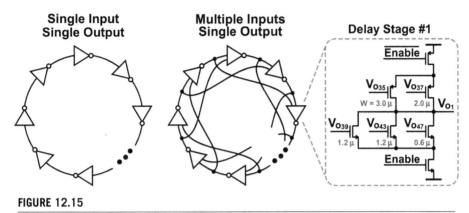

FIGURE 12.15

Reduction of effective delay per stage using multi-path oscillator implementation.

approach, each delay stage is connected to a simple register rather than a counter. These registers sample the transition states of the ring oscillator at the conclusion of each measurement, which yields the quantized phase state of the oscillator. One of the delay stages is also connected to a counter which keeps track of phase wrapping in the oscillator. By combining the phase state and wrapping information from the counter, the number of transitions of the oscillator within a given measurement window can be computed at much lower power dissipation than with the parallel counter method.

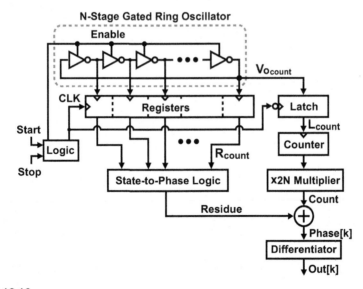

FIGURE 12.16

Phase sampling of GRO TDC state for reduced power dissipation.

Using the above techniques, a prototype GRO TDC was developed as shown in Figure 12.17 [9]. Here a 47-stage multi-path oscillator structure is utilized which achieves 6 ps delay per stage in 130 nm CMOS. The phase sampling operation is performed by seven interleaved channels, the total range input range is 12 ns, total active area is 0.04 mm^2, and power consumption varies from 2.2 to 21 mW depending on the input time between $Start(t)$ and $Stop(t)$ signals. Additional performance results for this structure are shown in the following sub-section.

12.2.3 **Measured results**

Figure 12.18 displays the measured output spectrum of the GRO TDC prototytpe shown in Figure 12.17 with 50 MHz sample clock, along with the time domain view of the output after 1 MHz lowpass filtering is applied. The measured output spectrum shows a tone corresponding to a 1.2 ps$_{pp}$ input signal, along with first-order shaping of the quantization noise. At lower frequencies, the noise profile resembles that of flicker noise, which is likely due to transistor device noise. The time domain view of the filtered TDC output clearly shows the periodic waveform of the 1.2 ps$_{pp}$ input signal, which verifies the effectiveness of scrambling and noise shaping since the 1.2 ps$_{pp}$ variation of the signal is significantly lower than the 6 ps raw quantization step size of the TDC. Overall, the effective resolution of the GRO TDC is 80 fs (rms) in 1 MHz bandwidth as measured by integrating the spectrum shown in Figure 12.18 across this bandwidth.

One issue of concern for the GRO TDC is the presence of dead zones. Figure 12.19 displays the filtered TDC output across a range of normal operation, but also shows a zoomed-in region that exhibits a dead zone of 1.1 ps in width. This dead zone is likely caused by partial loss of state, and therefore loss of scrambling behavior, in the GRO oscillator between measurements due to parasitic leakage paths which

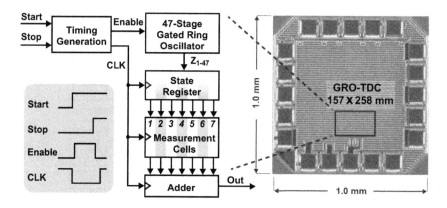

FIGURE 12.17

Example GRO implementation.

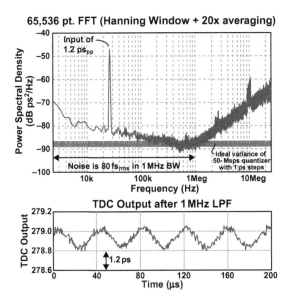

FIGURE 12.18

Measured output spectrum and filtered output of GRO implementation.

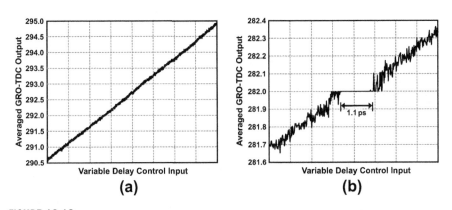

FIGURE 12.19

Filtered GRO output across (a) desired operating region, (b) deadzone region.

gradually discharge the oscillator capacitor voltages [10]. Fortunately, the multi-path GRO is free from such deadzones across much of its range, but this issue could be aggravated by higher leakage paths seen in advanced CMOS technology and/or slower clock rates for the GRO.

To better understand the impact of using the GRO TDC within a digital PLL, Figure 12.20 shows measured phase noise performance of a third order $\Delta\Sigma$

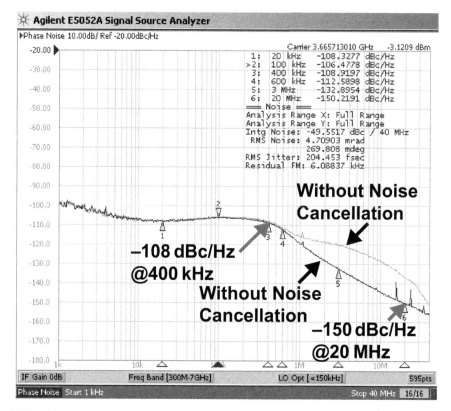

FIGURE 12.20

Measured phase noise performance of a digital fractional-*N* synthesizer utilizing the GRO TDC.

fractional-*N* synthesizer employing the above GRO prototype [11]. We see from this figure that the GRO TDC leads to excellent PLL phase noise of −108 dBc/Hz at 20 kHz offset frequency. In addition, the resolution of the GRO TDC enables techniques such as cancellation of the ΔΣ quantization noise produced by dithering of the divider value [11], which allows significant reduction of phase noise at higher offset frequencies as shown in the figure. Overall, the digital synthesizer prototype achieves less than 250 fs (rms) integrated jitter, which demonstrates the ability of the GRO TDC to enable high performance digital phase-locked loops.

As confirmed by the above-measured results, the GRO TDC offers significant advantages such as noise shaping of its quantization noise and mismatch and achieves a reasonably wide input range with relatively low area and power. High-performance digital PLL structures are enabled with this TDC topology. However, as with any technology, new approaches for improvement are always sought, and several examples of other noise shaping TDC topologies are discussed in the following section.

12.3 Advanced noise shaped TDC architectures

The noise shaping behavior demonstrated by the GRO TDC can be extended in various ways, including higher-order noise shaping and alternative implementations. Here we discuss recent architectural innovations which take the GRO concepts a step further and expand the field of noise shaping TDCs.

12.3.1 Higher-order noise shaping

Figure 12.21 displays an approach introduced in [12] to achieve a TDC implementation with higher-order noise shaping based on a MASH $\Delta\Sigma$ topology. Here a second-order noise shaping example is realized by cascading two first-order noise shaping GRO TDCs. While both TDCs use the same *Stop(t)* signal, the *Start1(t)* signal of the first TDC is set by the input, and the *Start2(t)* signal of the second TDC corresponds to the oscillator output of the first TDC. The key principle of operation is that the difference between the TDC inputs corresponds to the starting quantization error of the first TDC input, and this allows cancelation of the first-stage quantization error.

To better understand the TDC architecture in Figure 12.21, consider that the first TDC output can be expressed as

$$e_1[k] = in[k] + q[k] - q[k-1], \qquad (12.3)$$

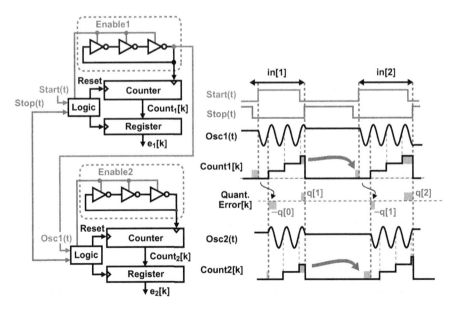

FIGURE 12.21

MASH TDC with second-order shaping.

where $q[k]$ is the quantization noise of the first TDC. The second TDC output can be expressed as

$$e_2[k] = in[k] - q[k-1] + r[k] - r[k-1], \qquad (12.4)$$

where $r[k]$ is the quantization noise of the second TDC. Let us now define $e_{\text{diff}}[k]$ as the subtraction of the two TDC outputs

$$e_{\text{diff}}[k] = e_1[k] - e_2[k] = q[k] - (r[k] - r[k-1]). \qquad (12.5)$$

Equation (12.5) reveals that $e_{\text{diff}}[k]$ yields the first-stage quantization noise accompanied by first-order noise shaping of the second-stage quantization noise. Finally, let us consider forming the overall output of the TDC, $out[k]$, as

$$out[k] = e_1[k] - (e_{\text{diff}}[k] - e_{\text{diff}}[k-1]) = in[k] + r[k] - 2r[k-1] + r[k-2]. \qquad (12.6)$$

Application of the Z transform of Eq. (12.6) yields

$$out(z) = in(z) + (1 - z^{-1})^2 r(z). \qquad (12.7)$$

Equation (12.7) reveals that second-order noise shaping is achieved for the overall TDC output defined in Eq. (12.6).

In addition to proposing the above approach for higher-order noise shaping, [12] demonstrated the concept with a third-order noise shaping TDC topology where the ring oscillator of the GRO TDC was replaced by a relaxation oscillator. The third-order noise shaping topology is achieved by cascading three rather than two TDC stages in a similar manner as the second-order example shown in Figure 12.21. As with any analog MASH $\Delta\Sigma$ circuit, one should note that matching between stages is a key issue in achieving the full potential of a higher-order noise shaping TDC [13].

12.3.2 Switched ring oscillator (SRO) TDC

As proposed in [14], one can extend the GRO TDC concept to the idea of switching between two frequencies as illustrated by the Switched Ring Oscillator (SRO) TDC topology shown in Figure 12.22. The GRO TDC can be thought of as an SRO TDC in which the lower frequency is zero.

The SRO TDC has the same property of performing first-order noise shaping of its quantization noise. The key idea is that the average frequency across a measurement interval varies between F_{low} and F_{high} according to the input time difference between the $Start(t)$ and $Stop(t)$ signals. An inherent first-order difference operation is performed by calculating the difference in count values, which is conceptually the same as performing a reset on the counter after every measurement interval as shown in the figure. This first-order difference operation, in turn, causes noise shaping of the quantization noise.

A key benefit of the SRO architecture is that it maintains transition movement within the ring oscillator stages such that leakage currents have a less disruptive

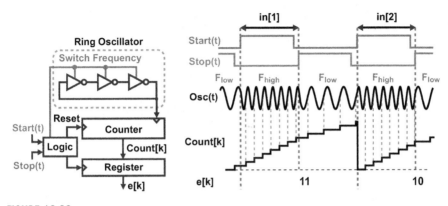

FIGURE 12.22

Switched ring oscillator (SRO) TDC concept.

impact compared to the GRO approach of gating off the oscillator between measurements. As such, dead zones in the SRO TDC are more easily avoided than with a GRO TDC [14]. However, a disadvantage of maintaining transition movement in the oscillator between measurements is that thermal and flicker noise of the oscillator devices may have a larger impact on the noise performance of the TDC. This issue will be discussed in more detail later in this chapter.

Another interesting implementation idea proposed in [14] is to form a pseudo-differential structure as conceptually shown in Figure 12.23. Here two SRO TDCs are utilized with switching control signals that are inverted with respect to each other. The average frequency of each TDC moves in an opposite direction between F_{low} and F_{high} according to the input time difference between the *Start(t)* and *Stop(t)* signals, and the overall output is formed as the difference between the individual TDC outputs. This pseudo-differential implementation reduces the impact of the TDC on generating supply noise since the average frequency of the combined TDCs remains constant, but carries the cost of increasing area and power by a factor of two. One should note that this same pseudo-differential implementation can be applied to GRO TDCs, as well.

12.3.3 **Other approaches**

Further TDC architecture variations can also be achieved by combining the GRO concept of noise shaping with other TDC techniques. In [15], two GRO TDCs are combined with a vernier technique to achieve improved resolution with the vernier approach in addition to the noise shaping behavior of the GRO. In [16], a two-step TDC using time amplification utilizes a GRO for its fine stage. Finally, in [17], an alternative approach to first-order noise shaping is proposed based on a "local oscillator"-based TDC. These various examples illustrate that noise shaped TDCs present a fertile research field that promises future advances in performance.

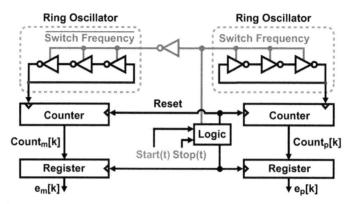

FIGURE 12.23

Psuedo-differential version of SRO TDC.

12.4 TDC implications for type II PLL topologies

While TDCs can be applied to a wide variety of PLL applications, it is worthwhile paying special consideration to Type II PLL topologies due to their prevalence. As shown in Figure 12.24, Type II digital PLLs contain an accumulator in the loop filter which forces the steady-state input into the loop filter, $e[k]$, to have an average value of zero. By purposefully introducing an offset, T_{offset}, before the loop filter, one can select the desired steady-state output value of the TDC by simply adjusting the value of T_{offset}. For a GRO TDC, this implies that the steady-state on-time of the TDC can be chosen as a system-level PLL design parameter as indicated in Figure 12.24.

Since the steady-state operating point of the TDC can be selected by design for a Type II PLL, the noise performance of the TDC can be optimized around that operating point rather than across its full range. In fact, in such cases, having a wide operating range for the TDC is primarily important for achieving fast settling behavior, and creative approaches can be taken to augment a limited-range TDC to deal with this issue [11]. An important observation for a GRO TDC is that reducing its on-time also reduces the impact of device noise on its performance since the current noise of the oscillator devices is dramatically reduced once the oscillator is gated off. Therefore, for best GRO TDC noise performance in a Type II PLL, one should set the offset as small as possible while still providing adequate margin in its range to track noise and perturbations of the PLL phase signal. For fractional-N PLLs, the phase variations due to divider dithering often constitute a significant portion of the required steady-state TDC range. For integer-N PLLs, a significantly smaller steady-state TDC range is required such that simpler TDC implementations can be utilized.

The above observations are helpful when comparing the GRO and SRO TDC structures. In the case of Type II PLL applications where a small steady-state TDC error can be chosen by design, the single-ended GRO TDC can offer improved noise

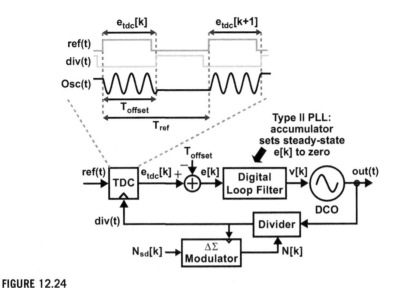

.**FIGURE 12.24**

Type II digital PLL and its impact on steady-state operating point of TDC.

performance compared to a pseudo-differential SRO TDC. To explain, the GRO approach of fully turning off the oscillator achieves lower device noise than allowing the oscillator to keep running between measurements, and a single-ended design avoids the pseudo-differential operation of always having one side at maximum frequency where substantial device noise occurs. The reduced steady-state operating range also alleviates the issue of deadzone behavior in the TDC since such regions are avoided by proper selection of T_{offset}. However, in the case of a Type I PLL and other applications where a large steady-state range must be supported, the pseudo-differential SRO TDC carries the advantage of having improved deadzone behavior with more uniform noise behavior than the single-ended GRO TDC.

CONCLUSION

The GRO TDC provides an elegant approach to achieving scrambled and first-order noise shaping of quantization noise and delay stage mismatch. Noise shaping is particularly beneficial for digital PLL architectures that inherently filter the TDC output. Further improvement in GRO resolution can be obtained using a multi-path implementation, and reduction in power consumption can be achieved using phase sampling with interleaved transition detection circuitry. Measured results of a GRO TDC prototype demonstrate excellent resolution of 80 fs (rms) in 1 MHz bandwidth, and enable a digital $\Delta\Sigma$ fractional-N synthesizer with less than 250 fs (rms) integrated jitter.

The GRO concepts can also be applied to a wide variety of other TDC architectures to achieve quantization noise shaping. Higher-order noise shaping is achieved with a MASH structure by cascading TDC stages which leverage the GRO concept of first-order noise shaping. Dead zone performance can be improved with an SRO TDC topology that maintains transitions in the oscillator between measurements. Finally, techniques such as vernier resolution enhancement, two-step quantization, and local oscillator-based structures complement the GRO concept and expand the horizons for achieving noise shaping TDCs with high resolution.

Acknowledgments

The author is grateful to Pavan Hanumolu, Ying Cao, and Matthew Straayer for their very valuable comments to improve this chapter.

References

[1] T. Rahkonen and J. Kostamovaara, The use of stabilized CMOS delay line for the digitization of short time intervals, *IEEE Journal of Solid-State Circuits*, vol. 28, no. 8, pp. 887–894, August 1993.

[2] S. Henzler et al., A local passive time interpolation concept for variation-tolerant high-resolution time-to-digital conversion, *IEEE Journal of Solid-State Circuits*, vol. 43, no. 7, pp. 1666–1676, July 2008.

[3] V. Ramakrishnan and P. T. Balsara, A wide-range, high-resolution, compact CMOS time to digital converter, *IEEE VLSI Design*, pp. 1–6, January 2006.

[4] A. Abas, A. Bystrov, D. Kinniment, and O. Maevsky, Time difference amplifier, *Electronic Letters*, vol. 38, no. 23, pp. 1437–1438, 2002.

[5] M. Lee and A. Abidi, A 9b, 1.25 ps resolution coarse-fine time-to-digital converter in 90 nm CMOS that amplifies a time residue, *IEEE Journal of Solid-State Circuits*, vol. 43, no. 4, pp. 769–777, April 2008.

[6] R. Baird and T. Fiez, Linearity enhancement of multibit deltasigma A/D and D/A converters using data weighted averaging, *IEEE Transactions on Circuits and Systems II: Analog and Digital Signal Processing*, vol. 42, no. 12, pp. 753–762, December 1995.

[7] S. J. Lee, B. Kim, and K. Lee, A novel high-speed ring oscillator for multiphase clock generation using negative skewed delay scheme, *IEEE Journal of Solid-State Circuits*, vol. 32, no. 2, pp. 289–291, February 1997.

[8] S. S. Mohan et al., Differential ring oscillators with multipath delay stages, in *Proceedings of the IEEE 2005 Custom Integrated Circuits Conference*, pp. 503–506, September 2005.

[9] M. Straayer and M. H. Perrott, A multi-path gated ring oscillator TDC with first-order noise shaping, *IEEE Journal of Solid-State Circuits*, vol. 44, no. 4, pp. 1089–1098, April 2009.

[10] M. Straayer, *Noise Shaping Techniques for Analog and Time to Digital Converters using Voltage Controlled Oscillators*, PhD Thesis, MIT, 2008.

[11] C.-M. Hsu, M. Straayer, and M. H. Perrott, A 3.6-GHz low-noise, wide-BW digital T-T fractional-*N* frequency synthesizer with a noise-shaping time-to-digital converter and digital quantization noise cancellation, *IEEE Journal of Solid-State Circuits*, vol. 43, no. 12, pp. 2776–2786, December 2008.

[12] Y. Cao, P. Leroux, W. De Cock, and M. Steyaert, A 1.7 mW 11b 1–1–1 MASH delta–sigma time-to-digital converter, in *IEEE International Solid-State Circuit Conference*, pp. 480–481, February 2011.

[13] S. Nortsworthy, R. Schreier, and G. Temes, *Delta-Sigma Data Converters: Theory, Design, and Simulation*. IEEE Press, 1997.

[14] A. Elshazly, S. Rao, B. Young, and P. Hanumolu, A 13b 315 fs (rms) 2 mW 500 MS/s 1 MHz bandwidth highly digital time-to-digital converter using switched ring oscillators, in *IEEE International Solid-State Circuit Conference*, pp. 464–465, February 2012.

[15] P. Lu, P. Andreani, and A. Liscidini, A 90 nm CMOS gated-ring-oscillator-based vernier time-to-digital converter for DPLLs, in *Proceedings of the ESSCIRC (ESSCIRC)*, pp. 459–462, September 2011.

[16] S.-H. Chung, K.-D. Hwang, W.-Y. Lee, and L.-S. Kim, A high resolution metastability-independent two-step gated ring oscillator TDC with enhanced noise shaping, in *Proceedings of 2010 IEEE International Symposium on Circuits and Systems (ISCAS)*, pp. 1300–1303, May 2010.

[17] F. Brandonisio and M. Kennedy, First order noise shaping local-oscillator-based time-to-digital converter, in *IEEE ICECS*, pp. 41–44, December 2010.

Index

Printed in the United States
By Bookmasters